DEPARTMENT OF MATHEMATICS
OBERLIN COLLEGE
OBERLIN, OHIO

The Shape of Content

The Shape of Content

Creative Writing in Mathematics and Science

edited by
Chandler Davis
Marjorie Wikler Senechal
Jan Zwicky

A K Peters, Ltd.
Wellesley, Massachusetts

Editorial, Sales, and Customer Service Office
A K Peters, Ltd.
888 Worcester Street, Suite 230
Wellesley, MA 02482
www.akpeters.com

Copyright © 2008 by A K Peters, Ltd.

All rights reserved. No part of the material protected by this copyright notice may be reproduced or utilized in any form, electronic or mechanical, including photocopying, recording, or by any information storage and retrieval system, without written permission from the copyright owner.

Library of Congress Cataloging-in-Publication Data

The shape of content : creative writing in mathematics and science / edited by Chandler Davis, Marjorie Wikler Senechal, and Jan Zwicky.
    p. cm.
Includes bibliographical references.
ISBN 978-1-56881-444-5 (alk. paper)
1. Mathematics--Literary collections. 2. Science--Literary collections. I. Davis, Chandler. II. Senechal, Marjorie. III. Zwicky, Jan, 1955-
PN6071.M3S53 2008
808.88--DC22   2008022399

Printed in India
12 11 10 09 08   10 9 8 7 6 5 4 3 2 1

Contents

Introduction	xi
Acknowledgments	xvii
Thanks	xix
Marco Abate (metafiction) *Évariste and Hélöise*	1
Colin Adams (humor essay) *Robbins v. New York*	17
Madhur Anand (poem) *Dissecting Daisy*	23
Sandy Bonny (short story) *Frames*	25
Wendy Brandts (prose poem) *Collisions*	35
S. Isabel Burgess (poems and prose poems) *Active Pass* *Magic Stretch Glove* *Tyndall Field* *Resonance* *There's something I need to say to you...*	37
Robin Chapman (poems) *Praying to the God of Leavetakings* *Brave New Biosphere* *The All of It*	45

Chandler Davis (poems) — 49
   Guided
   Our Study
   Presence
   Cold Comfort

Florin Diacu (nonfiction) — 55
   The Birth of Celestial Mechanics

Adam Dickinson (poems) — 77
   The Ghosts of Departed Quantities
   Eclipse
   Contributions to Geometry
   Great Chain of Being

Susan Elmslie (poems) — 87
   Algebra
   Chemistry

Claire Ferguson (essays) — 91
   Eine Kleine Rock Musik III
   Wild Singular Torus
   Igusa Conjecture

Emily Grosholz (poems) — 95
   Hourya
   Trying to Describe the Reals in Cambridge

Lauren Gunderson (short story) — 97
   The Ascending Life

Philip Holmes (poem sequence) — 109
   The Lines Remake the Places

Alex Kasman (short story) 113
    On the Quantum Theoretic Implications of Newton's Alchemy

Ellen Maddow (drama) 127
    Delicious Rivers

Marjorie Wikler Senechal (nonfiction) 139
    The Last Second Wrangler

Manil Suri (short story) 155
    The Tolman Trick

Randall Wedin (nonfiction) 175
    Breaking Down the Barriers

Paul Zimet (drama) 185
    Star Messengers

Biographical Notes 189

Introduction

The Shape of Content is an anthology of creative writing. The shapes are the literary forms of creative writing: poems, essays, drama, fiction, nonfiction. The content is broad too, but with a common thread: mathematical and scientific themes and sensibilities.

Creative writing about the content of mathematics and science is rare, and creative writing about the activity of mathematical and scientific creation is even rarer. And yet, when it occurs, it can be extremely popular, as well-known plays like *Proof* and *Copenhagen* and biographies like *A Beautiful Mind* and *The Man Who Loved Only Numbers* attest. What draws the public to these works? And why, given that something does, are there so few examples of literature that engages these themes? Mathematics and science are part of world culture, part of the human spirit, fit subjects for art of all kinds.

The paucity of art portraying them may reflect, partly, the genuine difficulty of conveying intellectual creativity of any kind. The creativity of mathematicians and theoretical scientists is especially hard to convey. We may not know what's whirling in the head of the composer, painter, writer, or experimenter we see furiously composing, painting, writing, or inventing, but we expect to be rewarded, by the end of the book, film, or play, with something we can hear, see, read, or use. A theoretician produces theory. The mathematical genius struggling alone in her attic with material so difficult almost no one else can understand it is a Romantic heroine par excellence: but how can a novelist or playwright bring that to life?

There is a precedent for our endeavor. Clifton Fadiman's charming anthology, *Fantasia Mathematica*, includes gems from Plato to Robert Heinlein, from Andrew Marvell to H. G. Wells. First published half a century ago, the book is still in print. But unfortunately, more for external than internal reasons—the pervasive anti-intellectualism in American life, the even more pervasive math phobia—*Fantasia Mathematica* is more museum than muse. Though widely admired among the cognoscenti, it has not stimulated the creation of a body of literature around mathematics and science that is badly needed. We hope *The Shape of Content* will help to do that.

This book arose from a unique, remarkable, and continuing effort to encourage practitioners who engage this content in their work—to give them opportunities to discuss important issues, to learn what others are doing—to encourage each other, to critique current work, to welcome young writers into the field, to spark collaborations, to forge networks and build a community. The story began eight or nine years ago when Robert Moody, one of Canada's leading mathematicians and a serious photographer, was invited to become the first scientific director of the new Banff International Research Station for Mathematical Innovation and Discovery (BIRS). BIRS's location, on the campus of the Banff Centre, Canada's globally acclaimed artistic, cultural, and educational institute, appealed to Moody. He hoped to link the creative activities of the two institutions.

With studios nestled in the woods, outstanding mentors, excellent performance spaces and a fine library, the Banff Centre nurtures aspiring, mid-career, and established musicians, visual artists, writers, dancers, and actors. The centre's world-class exhibitions, public readings, and performances enhance Banff's appeal to tourists year-round. BIRS, on the other hand, was inspired by an institute in the remote village of Oberwolfach, deep in the German Black Forest. Oberwolfach, as the Mathematisches Forschungsinstitut is usually called, is far from the tourist track. In spite of that, or rather because of that, it attracts the finest mathematical scientists in the world for weekly workshops year-round. At Oberwolfach, it's mathematics 24/7. (Oberwolfach is the setting for Manil Suri's short story, "The Tolman Trick," in this volume.)

BIRS and the Banff Centre could certainly coexist. But, with such different audiences and different rationales, could they also cooperate, beyond sharing a dining room, a fitness center, and hiking trails? In particular, could the close relations between mathematics and literature be explored together? Robert Moody raised this question in a leisurely conversation with Marjorie Senechal at an Oberwolfach workshop. From that conversation grew the idea of workshops on creative writing in mathematics and science, hosted by BIRS in cooperation with the Banff Centre's renowned Writing and Publishing Program.

Shaping Content

The first three workshops were held in 2003, 2004, and 2006. Marjorie recruited Chandler Davis as a co-organizer for the first two, and together they recruited Jan Zwicky for the third. We established unusual guidelines from the start.

First, the workshop participants would be a mix of mathematicians and scientists with active vocational interests in writing, and writers with active intellectual interests in science and mathematics. They would, we assumed, have a lot to learn from one another. Second, the workshops would be open to practitioners of all literary genres, since creative writing can be inspired by cross-genre insights. And third, organizers and participants would be on the same footing. Instead of the usual workshop model, with students and teachers, we all would submit original work for criticism and we all would criticize everyone else's.

Events proved us right on all counts. A poet helped a mathematician find a better way to tell the end of his story. A mathematician non-fiction writer helped a dramatist extend the ideas of her play, ideas a filmmaker sitting in on their discussions recast in light verse. The workshop format evolved; in the third year, we think, we got it right. Drafts were circulated early, two weeks if possible, but at least a day, in advance. We met for five forty-five minute sessions each day.

The sessions were varied. In some cases, they consisted of feedback on a proposal—reactions to the ordering of chapters, to the sketched content of chapters, sometimes to the shape of a whole manuscript. In one case, we spent the session putting together a reading list for a theatre project on the nature of time. In cases where we were presented with drafts of completed work, the sessions tended to focus on two elements, often simultaneously: the success of the piece as writing (often very detailed suggestions were offered) and the accuracy of the science or mathematics that was in or behind the piece. Writers found the presence of working mathematicians and scientists extremely helpful in this regard. Mathematicians and scientists found the presence of writers (perhaps especially poets, with their awareness of craft) illuminating. Cooperation with The Banff Centre, which helped adjudicate applications, was also critical here, because it brought to the mix of participants outstanding literary professionals who were able to assist the mathematicians in their writing far more effectively than they could have assisted one another. The third workshop culminated in a public reading, which you can hear on the Banff Centre's website, http://www.banffcentre.ca/programs/program.aspx?id=438.

Criticism at the workshops was often blunt and stringent, but was never offered, nor taken, personally. Was this just the result of a lucky mix of people? Not entirely. All of us were interested in art with a specific, often difficult, intellectual focus. The writers were not, could not possibly be, anti-intellectual; the mathematicians and scientists were not, could not possibly be, impervious to aesthetic questions. The gates were open to cross-disciplinary understanding. Moreover, in subtle ways that we explore below, we were engaged in a common enterprise.

Shape and Content

"I've seen that what's-it before... where have I seen that what's-it before?" muses a character in *Delicious Rivers*, a play by workshop participant Ellen Maddow (we include three scenes in this volume). "Where have I seen 'The Shape of Content' before?" you may be wondering. For indeed, we took our title from a small book by a great painter, printmaker, and photographer, Ben Shahn. Shahn's *The Shape of Content* is based on six Charles Eliot Norton lectures he gave at Harvard University in 1956.

The title appealed to us for two reasons. First, it states exactly what the workshops were about, and what this anthology is about. And second, as Shahn explores the mysteries of artistic creation in his deeply personal lectures, he grapples with questions faced by all creative artists, writers, mathematicians, and scientists as well as painters, sculptors, and musicians. The BIRS/Banff workshops "worked" not only because we shared, and shared alike, the common goal of giving shape to mathematical and scientific content, but because the creative challenges demanded by shape and by content are so much the same. "The genius so-called," writes Shahn, "is only that one who discerns the pattern of things within the confusion of details a little sooner than the average man."

Discerning pattern: that's what writers, mathematicians, and scientists do. Shahn's book can be read in many ways; it has *multiplicity*, as the Italian writer Italo Calvino would put it. "Form is not just the intention of content," says Shahn, "it is the embodiment of content. Form is based, first, upon a supposition, a theme. Form is, next, the relating of inner shapes to the outer limits, the initial establishing of harmonies. Form is, further, the abolishing of excessive materials, whatever material is extraneous to inner harmony, to the order of shapes now established. Form is thus a discipline, an ordering, according to the needs of content."

Nineteen years after Shahn gave his Norton lectures, Italo Calvino was invited to speak in that series. He died on the eve of his departure for Harvard. His widow found the first five lectures, neatly typed, in his briefcase; for the sixth, she found only the title. The lectures were published posthumously, in *Six Memos for the Next Millennium*. Calvino had worked on the lectures for over a year, Esther Calvino wrote in her introduction to this slim volume. He "stood before the vast range of possibilities open to him and he worried, believing as he did in the importance of constraints, until the day he settled on a scheme to organize the lectures." *Lightness. Quickness. Exactitude. Visibility. Multiplicity. Consistency.* "I would like to devote these lectures to certain values," he wrote, "qualities, or peculiarities of literature that are very close to my heart, trying to situate them within the perspective to the new millennium."

Qualities, peculiarities of *literature?* Calvino's qualities are very like Shahn's! The embodiment of content is visible. A supposition, or theme, supplies quickness. Harmony is consistency. Abolishing excessive materials yields lightness. And discipline is exactitude.

The mathematical genius struggling alone in her attic is grappling with these qualities too. She seeks not *a* proof of her theorem, but *the* proof. Paul Erdős, the great twentieth-century mathematician who loved only numbers, an atheist, claimed that God has a book in which the best proof of every theorem is written. Erdős never listed the criteria a proof must satisfy to be inscribed in God's book: he didn't need to. Though no one has seen the book or ever will, all mathematicians know that Euclid's proof of the infinitude of primes is in it, and no mathematician doubts that computer-generated proofs, the kind that methodically check case after case, are not. The proofs in God's book are elegant. They surprise. In other words, they are light, quick, exact, and visible. They are metaphorical; they admit multiple interpretations. They are consistent, a new brick on the pyramid of mathematical knowledge constructed over the millennia by all the cultures of the world.

For our *The Shape of Content*, we solicited contributions from participants in all three workshops and selected those that best conveyed the spirit, the meaning, and the achievements of the series. The range in style, tone, expertise, genre and subject matter provides, we hope, a snapshot—live gestures caught mid-air—not a finished portrait. We hope this collection will attract readers who will become the writers and editors of its sequels.

Chandler Davis
Marjorie Senechal
Jan Zwicky

Acknowledgments

Robin Chapman's "Brave New Biosphere" originally appeared in *Free Verse*. "Praying to the God of Leavetakings" originally appeared in *Nimrod* (Vol. 51, No. 2, Spring/Summer 2008).

Adam Dickinson's poems "Great Chain of Being," "Contributions to Geometry: The Gulf Stream," and "Eclipse" were published in *Kingdom, Phylum* (Brick Books, 2002); "The Ghosts of Departed Quantities" was published in *Problematical Recreations* (Littlefishcartpress, 2008). Reprinted with thanks.

Sue Elmslie's poems "Algebra" and "Chemistry" were both published in her collection, *I, Nadja, and Other Poems* (Brick, 2006, www.brickbooks.ca).

Claire Ferguson's essays "Eine Kleine Rock Musik III," "Wild Singular Torus," and "Igusa Conjecture" first appeared in *Helaman Ferguson, Mathematics in Stone and Bronze* (Meridian Creative Group, 1994); the accompanying photographs also appeared in that book.

Emily Grosholz's "Trying to Describe the Reals" first appeared in her book of poems, *The Abacus of Years* (David R. Godine, 2002). "Hourya" was first published in *The Hudson Review* (Vol. LXI, No. 1, Spring 2008).

Alex Kasman's story "On the Quantum Implications of Newton's Alchemy" appeared in *Analog*, 2007.

Ellen Maddow's play *Delicious Rivers* was developed, in part, in the BIRS/Banff creative writing workshops. It was first performed at La MaMa in New York in 2006.

Manil Suri's short story, "The Tolman Trick," was first published in *Subtropics* and is reprinted by permission of Manil Suri and Aragi Inc.

Paul Zimet's play *Star Messengers* was commissioned by the Kahn Liberal Arts Institute at Smith College and was first performed at Smith in April 2001.

Thanks

The editors would like to thank Carol Holmes, Kim Mayberry, and Robert Moody for their generous cooperation in planning and executing the workshops at BIRS/Banff.

Wendy Brandts dedicates "Collisions" to the memory of Lynn E. H. Trainor, her Ph.D. supervisor, with gratitude for his inspired guidance.

Lauren Gunderson dedicates her story, "Ascending Life," to her grandmothers, Irene and Beatrice.

Marjorie Wikler Senechal is pleased to thank the Sophia Smith Collection at Smith College for permission to quote from the Wrinch Papers, and Reading University for permission to quote from its archives.

Marco Abate

Évariste and Héloïse
A Wish List for a Story

A few months ago I was asked to propose a story about Évariste Galois, the French mathematician, for the French comic books market. I decided from the very beginning that I didn't want it to be a biographical sketch; I wanted to use the (admittedly interesting) historical character to do something else, something closer to my own personal obsessions. So I started playing around with a few tales I'd heard, a few people I'd met, a few books I'd read... And Héloïse was born, asking for her story to be told. And Auguste Dupin, the original Poe detective, was living more or less at the same time as Galois, a coincidence too good to let pass. And I was also sort of thinking (well, fiddling around might be a better expression for what I was actually doing) about possible relationships between magic and mathematics, and their different but sometimes uncannily similar ways of the exploring unknown world...

Anyway, given my writing habits, the final product will probably be completely different from what is contained in these few pages; any suggestions for directions I might pursue (or ought not to pursue) will be appreciated, carefully considered, and then thrown away together with the rest of the story when the editor discovers that I'm writing something completely different from what he had in mind when he asked me in the first place. Oh well...

Enough chatter. Let us start by talking a bit about Galois, and the reasons I'm interested in him.

> GALOIS, ÉVARISTE (1811–1832) French mathematician famous for his contributions to higher algebra, gave his name to the Galois theory of groups. He was born Oct. 25, 1811, at Bourg-la-Reine, where his father was mayor, and entered the Lycée Louis-le-Grand in Paris in 1823. Despite his genius for mathematics he was evidently a difficult student. He failed

twice in the entrance examinations to École Polytechnique; he was accepted at the École Normale in 1830 but expelled the same year for a newspaper letter about the actions of the director during the July Revolution. In 1831 he was arrested for a threatening speech against King Louis Philippe, but acquitted; then shortly afterward he was sentenced to six months in jail for illegally wearing a uniform and carrying weapons. He died on May 31, 1832, when only 20 years old, from wounds received in a duel, possibly with an *agent provocateur* of the police.

Galois published only a few small papers; three larger works on the theory of equations were refused by the French Academy. Knowledge of his mathematical achievements stems mainly from a letter to his friend Auguste Chevalier written on the eve of his fatal duel and from posthumous manuscripts.

(Encyclopædia Britannica, Vol. 9, p. 989, 1962 edition)

(More on this fateful letter later.) Even if he probably won't admit it, Galois is a teenager through and through. A genial teenager, no doubt, but he has all the stubbornness and passion typical of teenagers, only more so. If he gets into something he goes all the way, completely. No stopping halfway, no shades of gray for him. He has no doubts about himself, about being right in whatever he is doing, mathematics or politics or whatever else strikes his fancy (ok, not much else strikes his fancy, but you know what I mean). He is very passionate, very vocal about his opinions and beliefs, and he lives in a period where being passionate and vocal is both dangerous and exciting. He has the feeling that he is finally solving a problem that eluded the best mathematicians for centuries, and at the same time he believes that his political actions can make a difference to the world—the dream of any teenager. And here lies the first conflict: the world is not paying any (well, hardly any) attention to him. His papers are not recognized: they are lost, or, even if published, fail to get him the attention he craves. He firmly believes in the necessity (and the coming) of a new French Revolution in the name of the Republic; but his efforts produce no results, he is only one of the many young people in one of the many more-or-less-secret societies sprouting up from nowhere in those years. And he doesn't even manage to get recognized as an important member of any of these societies. All his genius, his passion, his efforts—for nothing. Doing mathematics is not a relief for him, it is a necessity. It is both a goal on its own and a means to be recognized by the world. The conflict lies in the distance between the energy he spends and the little he achieves.

There is another aspect of his character that I find interesting: he apparently has no sex life. And this is not standard; for a young man of his age and condition going with prostitutes was absolutely normal. Apparently, he transmutes all his sexual energy (which he certainly has a lot of) into mathematics and politics. In particular, he has a Dionysian approach to mathematics; he feels his work as a physical, sensual struggle against logical difficulties that are very concrete to him. He is no Apollonian thinker tinkering with abstract nonsense; the mathematical world he inhabits is as concrete, as corporeal, as alive as the so-called real world. There is a parallel with his political choices, too. The Saint-Simonian secret society (advocating the government by scientists in the name of reason for the welfare of the people) that he initially associates with is too moderate for him; after a short period he moves to more extremist republican secret societies. He is more interested in doing something for the republic right now, without reflecting too much on what happens next. Even though I do not want to stretch too much the analogies between his mathematical attitudes and his political attitudes, one thing is clear: he has a fully passional, emotional, sensual approach to all important aspects of his life—but sex.

Auguste Dupin, on the other hand, is the embodiment of the Apollonian thinker: no emotion, only pure logic, pure analysis. Which does not mean he does not enjoy using his abilities; on the contrary, he enjoys them exactly because he is not emotionally affected by them (or, better, he affects not being affected).

> As the strong man exults in his physical ability, delighting in such exercises as call his muscles into action, so glories the analyst in that moral activity which disentangles. He derives pleasure from even the most trivial occupations bringing his talents into play. He is fond of enigmas, of conundrunums, of hieroglyphics; exhibiting in his solutions of each a degree of acumen which appears to the ordinary apprehension preternatural. (...) At such times I could not help remarking and

admiring (although from his rich ideality I had been prepared to expect it) a peculiar analytic ability in Dupin. He seemed, too, to take an eager delight in its exercise—if not exactly in its display—and did not hesitate to confess the pleasure thus derived. (...) "But it is by these deviations from the plane of the ordinary, that reason feels its way, if at all, in its search for the true. In investigations such as we are now pursuing, it should not be so much asked 'what has occurred,' as 'what has occurred that has never occurred before.' (...) I said 'legitimate deductions;' but my meaning is not fully expressed. I designed to imply that the deductions are the sole proper ones, and that the suspicion arises inevitably from them as the single result. (...) Now, brought to this conclusion in so unequivocal a manner as we are, it is not our part, as reasoners, to reject it on account of apparent impossibilities. It is only left for us to prove that these apparent 'impossibilities' are, in reality, not such."

(Edgar Allan Poe, The Murders in the Rue Morgue, 1841)

(More than a shade of Sherlock Holmes here.) The world has to be stripped down to its basic elements, and reconstructed starting from there. He is rationality's voice (not to be confused with the voice of reason). He is unaffected by emotion; Galois is pure emotion. Galois is clearly unbalanced; Dupin never falters. Galois has a fully sensual approach to an ethereal subject such as mathematics; Dupin has a purely logical approach to an earthly subject such as murders. A few years older than Galois, Dupin meets him for the first time at a Saint-Simonian meeting (Dupin would have been a Saint-Simonian in his youth, no doubt). Dupin appreciates the vitality of Galois' intelligence, Galois needs the moderation of Dupin's ways; they become friends (Fig. 1).

And then, there are the dreams. Or visions. Most mathematicians, when working, live in a world of their own; Galois (or, at least, my fictional Galois) goes a step further. He actually sees his mathematical world; he enters it. In the dreams/visions he has when working (and not only then) he is able to interact physically with this ideal universe, this universe of ideas. It is (or, at least, he perceives it as) a physical universe, with mass and substance and winds and storms. It has lands and landscapes, boulders

Fig. 1. Galois and Dupin. (Copyright 2005 Paolo Bisi.)

and crevasses, infinite stretches of ugly trees ending just beyond the horizon and isolated phosphorescent flowers imploring wanderers to care for them forever. It is shaped by his research, in ways he cannot predict. It directs his own discoveries, in ways he couldn't have imagined.

What I have in mind is a sort of crossing between the platonic world of ideas we mathematicians might be exploring, and the "ideaspace" some philosophers (and some magicians) wonder about, a dimension inhabited by all the concepts humans (and aliens) could imagine, where all stories are true. It exists outside our limited notions of time and space, as fully given as Einstein's relativistic four-dimensional space is. A portion of it is devoted to mathematical ideas, and the strength of the sensual approach of Galois to mathematics can take him there.

An important point here is that, even if he has visions, Galois is not crazy. He behaves in all respects as a sane person, a passionate, instinctive, unrestrained basically sane person. He just happens to see something that others don't, exactly as others don't see (or understand) his mathematics. He will endure a lot of stress during the story, and at the end the boundaries between what is real and what is not will be a lot more blurred than at the beginning, but this is in the nature of the story and has nothing to do with his sanity. Galois and Héloïse are not, and won't become, crazy, visions notwithstanding.

In the story I do not want to dwell too much on the philosophical aspects behind this idea of "ideaspace", even though I'm sure Dupin will have something to say about it. I need it and it suits me for four different reasons. First, it gives me a way of visualizing in a metaphorical way both the mathematics Galois is doing and the way he is doing it. I imagine a sort of Fomenko landscape, impossible and horrifying and yet deeply beautiful (Fig. 2). It looks hellish to us because we are not equipped to interpret it, as Euclidean beings immersed in a highly non-Euclidean space. It causes sensory overloads; and yet, we glimpse its inner beauty, its hidden symmetries. It might be a sort of 3-dimensional projection of a much higher-dimensional algebraic manifold (like the way aperiodic tilings are obtained as projections of a more regular 5-dimensional grid), a manifold whose symmetries are given by the Galois group of some equation.

Fig. 2. Galois in the ideaspace. (Copyright 2005 Paolo Bisi.)

Second, the horrifying aspects of Galois' dreams fit very well with the dreadful backdrop of Paris during the cholera epidemic of 1832 (almost 20,000 deaths in less than five months). They can be deeply related, both visually and metaphorically. In the last part of the story, Galois will go from visions to reality and back almost seamlessly. From the incomprehen-

Fig. 3. The cholera epidemic. (Copyright 2005 Paolo Bisi.)

Marco Abate: Évariste and Héloïse

sible horrors of bodies left dead in the gutters of everyday Paris (Fig. 3) to the horrors of outlandish landscapes, incomprehensible and yet carrying the promise of being tamed —one day. I don't yet know precisely how we'll get there, but I do know that while lying in the grass waiting to die after the fatal duel, Galois will be in his visions, where he at last... well, I'll tell you later what at last.

Third, the surface detective story will also deal with the ideaspace. (I need a surface story to keep casual readers interested. The reason I'd like to tell this story is because of the characters involved; the reason most readers will—I hope—buy it is because of its surface.) One of the elements leading to the fatal duel will be a sub-sect of Saint-Simonians meddling with magic and architecture, trying to get a hold on the ideaspace. In a completely scientific way, of course. It is not as wacky as it sounds; much wackier ideas have been "scientifically" pursued in the eighteenth century (and they still are even in this century, for that matter).

Fourth, it will give me a way to let Galois and Héloïse interact. Because Galois is not alone in his visions. There is a woman. He doesn't know who she is (she will eventually know who he is, but more about this later). She shouldn't be there. She is distracting. She is enticing. She cannot be explained away. She is affecting his visions, making them... better? She is frightened, at least at the beginning. But then, slowly, they start to interact, to communicate. And the ideaspace changes with them. It is like a dance, a courting of ideas and souls, modifying their mindscape. He wants her to be there, he longs for her. He will take crucial decisions in crucial moments because of her; and she will learn a great deal from him. And in the end... well, I'll tell you later what will happen in the end.

I'll come back to Héloïse soon. But first I have to give a very rough synopsis of Galois' half of the story. The starting point is the (historical) suicide of Nicholas, Galois' father, in his Paris apartment. Apparently, Galois' father killed himself because of a libel insulting his friends and relatives. The libel appeared under his name, while in reality it was written by a Jesuit who, for political reasons, wanted him to resign as mayor (I'm making a long story short). Galois, who loved his father dearly, is very much affected by his death; he cannot understand how could he kill himself. And indeed Dupin's position is that he didn't; there is no logical reason for it, the libel is too weak an excuse. The only logical possibility is that he was killed, and they (Dupin and Galois) have the responsibility to discover why, and by whom.

So the idea is to structure Galois' half of the story as an investigation, which also somehow parallels his mathematical investigations. Galois and Dupin will discover that the building where Galois' father lived in Paris was built according to particular mystical proportions and symmetries (which neatly fit with Galois' discoveries about the symmetry group of an equation, and thus with the visual shape of the ideaspace he is traveling in) by a sub-sect of Saint-Simonians, who wanted to use the building as a starting point for the investigation of the ideaspace. To get into the ideaspace, they need to be in a very particular frame of mind; and to reach it, they are ready to use any possible means, starting from the shape of the building up to morbid rituals involving recently deceased Parisian citizens. Of course, they are doing all of this for the eventual benefit of humanity—and because they want to know what is there. (An important and difficult point here is that I don't want them to be "bad guys"; they're doing what they're doing because they think it is the right thing to do. There are often very good reasons for doing opposite, and incompatible, things, and every choice is wrong—or right, depending on which side of the mirror you currently stand.)

Dupin's and Galois' conjecture is that Galois' father was killed because he got in the way of the Saint-Simonians' goals. So Galois wants to try and actually get in their way too, and this will lead to his imprisonment and eventually to the duel.

We follow Galois' investigations, prompted and guided by Dupin. The visions intensify, as well as Galois' political involvement. All his actions become more intense, more passionate—and his frustration grows too. In a meeting of ardent republicans, he proposes his by now (in)-famous toast "To the king!", glass in one hand, dagger in the other. The next day he is arrested for defaming the king, his name suggested to the authorities by the Saint-Simonian sub-sect, who didn't appreciate his meddling.

In prison. A living nightmare. The horrifying beauty of his visions his only solace. She is there, too, in his visions. Dupin keeps investigating. Cholera rages throughout Paris. Unexpectedly, this helps Galois: he is released because of city health reasons. He comes out devastated. He keeps going because of his mathematics—and because of her. But she lives in his visions only. He confronts the Saint-Simonian sect, and their reaction leads to the duel. His visions merge with the hellish Parisian alleys ravaged by cholera.

The night before the duel, he writes a few letters.

> All night long he had spent the fleeting hours feverishly dashing off his scientific last will and testament, writing against time to glean a few of the great things in his teeming mind before the death which he saw could overtake him. Time after time he broke off to scribble in the margin "I have not time; I have not time," and passed on to the next frantically scrawled outline. What he wrote in those last desperate hours before the dawn will keep generations of mathematicians busy for hundreds of years. He had found, once and for all, the true solution of a riddle which had tormented mathematicians for centuries: under what conditions can an equation be solved?
>
> (E. T. Bell, Men of Mathematics, 1937)

This famous description is historically inaccurate. Galois actually wrote just a résumé of his previous work, with some new remarks and further explanations (and indications of missing details that he had "not the time" to fill in), but the main mathematical core of his researches was contained in the memoirs he had already submitted to the French academy (where they were either rejected or lost). He also wrote a couple of short letters to friends, briefly explaining the reasons for the duel. (Which, in reality, had nothing to do with politics or the Saint-Simonians, but was apparently due to some quarrel over a girl: his first approach to sex led him to death. How poignant.)

But here we are not interested (that much) in the historical Galois. We can however use this feverish sleepless night to bring him to the final merging of his own two realities. In an abandoned field at the outskirts of Paris, alone; and in moving waves of unfathomable symmetries, with her. Dupin arrives too late to save him, to tell him how he found out beyond any shadow of a doubt that Nicholas Galois killed himself, possibly because the peculiar structure of the building he lived in had a magnifying effect on his depression. It was no direct fault of the Saint-Simonian subsect. The suffering, the prison, the duel: it has all been for nothing. (Or has it?)

Fig. 4. Héloïse. (Copyright 2005 Paolo Bisi.)

Galois' story will be intertwined with Héloïse's story, in such a way that main events in one story will be reflected (distorted, magnified, symmetrized—if such a word actually exists) in the other. Héloïse (Fig. 4) is about thirty years old, a high-school mathematics teacher in modern-day Paris.

Yes, I know Héloïse well, nobody better than me, I think. We've been best friends since first grade... A funny little mouse she was—and then look at the beauty she became! But she always kept her mousy ways, somehow... mostly just doing what was expected of her. Even though stubborn to the bone sometimes she was, mind you, when she really wanted something there was no keeping her back. Look at the way she got François. All us girls drooling after him, and who got the prize? Mousy Lise! Unassuming, unpretending Mousy Lise... but she had more than good looks on her side. And I'm not only talking about her brains, even though she had a lot of it, and François had always admired that. She was, you know, a bit wild in bed. Just an eensy-teensy bit wild, François wouldn't have liked real wild, just enough to get him hooked. And hooked he was! He's always been faithful to her since then, and he could have had all the girls he wanted. Five years older than us, good looks, very good prospects before and that incredible job of his after...

Maybe it was that freak streak in her. Everybody's best girl, respectful, studious, polite, never a worry for her parents or teachers... and then bang! She went and did something completely outrageous, like that, out of the blue. And she was awfully proud of that, whatever that was, ready to defend it against the whole world. You know that when she was sixteen she got a piercing on her left nipple? Incredible. And her mathematics! She wanted to be a mathematician, for heavens' sake! A mathematician! Can you believe that? Not even François could dissuade her—well, eventually he did, but you know what I mean, their daughter wasn't exactly what you'd call expected... She had to cope with that, and she did, you have to grant her that. François was wonderful; they got married right away, and then he bought that wonderful apartment... he is a big shot now, you know, in that financial firm, what's-its-name... a lot of work, but he is good, I can tell you. Why Héloïse wanted to go teaching is over my head. With her house, and her husband, I'd just stay home the whole day... and night... It must have been that freak streak in her. But she is a very good mother, I grant her that, and they tell me she is a good teacher too.

No, I don't see her that much anymore. With her work, and her daughter, and her house, and her husband, she has no time for her old friends, or at least so she says... Happy? Is she happy? What kind of question is that? Of course, she is! I would like to have her house, and her husband! Look at the place I am forced to live now; this is not what I expected when I met Henri... Héloïse got François first, and Henri was his best friend then... She got the better deal, she did, don't get me started about Henri, you'd better not... She'd better be happy! Don't you think so?

(Excerpts of an interview with Denise Marchand, Ici Paris, 2004)

Héloïse lives, together with her husband and her six-year-old daughter, in the same apartment Galois' father lived in. She wanted to become a professional mathematician, and she was quite good at it, but during grad school she got pregnant, and you know how these things go. She found herself at home caring for the baby while her soon-to-be husband worked his way up in a financial firm. As soon as she could, she got a temporary position as a high-school teacher, but it is not what she had hoped to become. Her life is shades of gray. She is a good teacher, she is a good mother, she tries to be a good wife—and this leaves no time for her own needs and desires. Which is something her husband doesn't understand, doesn't even imagine could be a problem. In his own terms, he thinks he is a good husband; he just doesn't get it.

Héloïse (possibly because of the building they live in) has a recurrent dream; not often, but often enough. She dreams of another, incoherent, formless, horrible yet beautiful place. With possibly somebody else there. A shadow. An incoherent formless alluring shadow.

This is her life: her daughter, her work, her house, her husband, sometimes the dreams. And then an old friend, Antoine, shows up.

> CHEVALIER, ANTOINE (1973) Born in Toulouse on May 4, 1973, of Louis Chevalier, journalist, and Marie Lagrange, secretary in a law firm. A brilliant student, he was accepted at the École Normale in 1991. There, he received his doctorate in Mathematics in 1998, under the direction of Prof. François Berteloot. His ground-breaking work on the application of novel algebraic techniques to the study of holomorphic dynamical systems won him a postdoctoral fellowship at the University of California, Berkeley, for the years 1999–2000, and then a four-year position as assistant professor at Princeton University. In 2005 he accepted a permanent associate professorship at the École Normale Superiéure in Paris; he is now one of the youngest professors at the Institut. Besides doing mathematics, he likes walking the streets of Paris at night, and folding unreal origami shapes.
>
> *(American Mathematical Monthly, Vol. 127, 2005, p. 327)*

Antoine and Héloïse were grad students together, then she got pregnant, his postdoc took him to the States, and they lost contact. After a few years abroad, he landed the position in Paris, and now here he is. They meet, they start talking, and it is like the most natural thing, as though time hadn't passed. He inquires about her mathematics (which had something to do with Galois theory); she was doing quite good work at the time, and since then nobody else has gone in the same direction which now looks even more promising than before, why doesn't she give it another try? She must complete her work; she can do it!

She hesitates, but eventually she decides to try. And as soon as she re-enters mathematics, she realizes how much she has missed it. To be in a world she feels as her own, where success can be achieved, if ever, by her own efforts only, where she doesn't depend on anybody and nobody depends on her… But doing mathematics requires time, time that must be taken off the rest of her life. Her husband doesn't understand, he feels rejected, he feels jealous—their sex life suffers.

At the same time, the dreams become stronger, more structured. She is afraid, and attracted. She is no fool; she soon understands the connection between coming back to mathematics and the dreams. The formless presence she sensed before becomes more and more concrete, and finally she recognizes him: Galois. They do not talk, in the dreams; they communicate by manipulating the landscape in ways they later translate into mathematics (and ideas, and emotions). She knows his story (all mathematicians do), and she knows he is going to die soon. Or is he already dead? After all, it is only a dream, isn't it? It just doesn't feel like a dream anymore.

> One way to look at rationality in dreams is to classify different levels of lucidity. At the highest level, the dreamer would not only be aware of dreaming, but also possess complete understanding of the implications of this knowledge, and would behave in accordance with that understanding on all levels from thought to action. The lowest, minimal level of lucidity would be realization of dreaming, but without understanding how dreaming is different from waking, and without acting on the lucidity at all, mistaking events, characters and consequences with those from waking life. Yet, degrees of rationality vary

from moment to moment in dreams, so that one wishing to use a scale of levels of lucidity would have to rate each decision, action, or response of the dreamer independently. Averaging the lucidity levels in a dream might be a way of establishing a lucidity "score" for the dream. All of this is for future research to decide.

(Lynne Levitan, A Fool's Guide to Lucid Dreaming, NIGHTLIGHT 6(2), Summer, 1994)

Héloïse accepts the dreams, as she has accepted all that has happened to her in her life: her daughter, her husband's decisions, Antoine's suggestion of going back to mathematics… It's one of the reasons I think Galois and Héloïse complement one another so well. Galois is all action, he tries as much as he can to steer his life in the direction he wants. Héloïse, instead, is led by her life; she tries to make the best of what happens, but she almost never tries to make things go in the direction she'd like them to go. On the other hand, Galois is unable to organize his life and his efforts in a productive way, while Héloïse knows all about organizing things (as almost any woman who works and is also a mother and a wife does).

She cannot talk with her husband about the mathematics, or about the dreams either. They become more and more estranged, even though on the surface apparently nothing has changed. And then it happens. The husband is away, Antoine is there, they start talking, drinking, they become intimate, they end up in bed.

She finds in him something that up to that moment she almost didn't know was missing. Her mathematics, her lover… she now has a life of her own, it doesn't matter what is going to happen, it doesn't matter how difficult it is, this is what she needs now. Even though managing her life has become a nightmare. Keeping all the elements of her life together and separated at the same time is almost impossible, but she doesn't know how (and doesn't want) to leave anything behind.

Then, possibly because of the properties of the building, her husband discovers something—and a crisis is precipitated. He doesn't want to lose her (or to give her more space either—she's always had, I've always given her, enough space, haven't I? I didn't do anything to deserve this!). He just wants things as they were. He throws away her mathematics. He doesn't know what he did wrong—he honestly tries to understand, but he cannot.

Again, he is not a bad person, he is trying to do his best; but sometimes the best leads in the wrong direction. And I don't want to be preachy; I hope to portray them as realistically as I can, with their strengths and their weaknesses, and not as role models to follow or despise.

In a particularly harsh row, he hits her—for the first and only time. She falls down, loses consciousness, and finds herself in the dream. Galois is there; he is dying, hit by a bullet in the chest, the morning of the duel. She approaches him; they touch. They embrace, they make love. Giving, at last, to somebody who understands. Content, in a moment they both want to be in. She gives him recognition at last, in the fully corporeal instinctive and fulfilling way that only a woman can give to a man; and he gives her the strength to steer herself at last (and no, I'm not going to say "as only a man can give to a woman", because first it is not true, and second, one should not stretch one's rhetorical arguments too far; but Galois is the right person to give her that strength in that precise moment). At the end of his life, Galois finally obtains what he has most desired, and he gives her what she has most needed then. Maybe it is a vision, or just a dream, or maybe it is actually real, in an outlandish Fomenko-Little Nemo way, it doesn't matter; his life ended as he wanted it to, and her life started again, as she hadn't let herself imagine it could.

She awakens in a hospital: head concussion, nothing serious. When she leaves the hospital, she leaves her husband too, taking her daughter with her. She doesn't move in with Antoine, at least not yet. She wants to be in charge of her life, now; she's dreamed of it long enough. And she leaves something else in the hospital; she consciously decides to. She doesn't need it anymore. Her piercing. It was on her left nipple. Over her heart. In the shape of a small, but perfectly recognizable, bullet.

THE END

Colin Adams

Robbins v. New York

The question before the state's highest court, the Court of Appeals, was whether a man named James Robbins was guilty of selling drugs within 1,000 feet of a school—which carries a longer sentence—when he was arrested in March 2002 on the corner of Eighth Avenue and 40th Street in Manhattan and charged with selling drugs to an undercover police officer.

The nearest school, Holy Cross, is on 43rd Street between Eighth and Ninth Avenues. How to measure? On foot, Mr. Robbins's lawyers argued, the school is more than 1,000 feet away from the site of the arrest, because the shortest route is blocked by buildings. But as the crow flies, the authorities said, it is less than 1,000 feet away.

Law enforcement officials calculated the straight-line distance using the Pythagorean theorem ($a^2 + b^2 = c^2$) measuring the distance up Eighth Avenue (764 feet) as one side of a right triangle, and the distance to the church along 43rd Street (490 feet) as another, to find that the length of the hypotenuse was —907.63 feet.

Lawyers for Mr. Robbins argued that the distance should be measured as a person would walk it because "crows do not sell drugs." But in a unanimous ruling, the seven-member Court of Appeals upheld his conviction and held that the distance in such cases should be measured as the crow flies.

—Michael Cooper, *New York Times*, Nov. 23, 2005

Today, the Supreme Court heard oral arguments in the case of *Robbins v. New York*. Jane Hausdorff represented the State of New York before the court.

Hausdorff: The nub of the question is how the legislators intended us to measure distance. Did they intend the Euclidean metric, which measures distance in a straight line from point A to point B, often described as "how the crow flies," or were they intending the taxicab metric, also known as the Manhattan metric, which measures the total distance traveled as you walk from point A to point B if you are required to stay on the streets that meet at right angles and are not allowed to take the diagonal?

Justice Thomas: Couldn't they have intended some other metric all together?

Hausdorff: I'm sorry, you honor. I'm not sure what you mean.

Justice Thomas: For all we know, they intended the Weil-Peterson metric.

Justice Ginsburg: Clarence, I believe the Weil-Peterson metric only applies to Teichmueller space.

Justice Thomas: Okay then, how about the Hermitian Yang-Mills-Higgs metric?

Justice Breyer: That applies to complete Kahler manifolds. Can we stick to metrics for measuring distance in the plane?

Justice Souter: What if the defendant were selling drugs on a subway train that was passing beneath the ground near the school? How would the distance be calculated then?

Hausdorff: Well, you honor, I guess it should also be measured as the straight line distance.

Justice Souter: But a crow can't fly through dirt. If this is "as the crow flies," then the crow would be forced to fly several blocks out of the way, down the subway entrance, through the subway station and then down the tunnel to the point of sale. That would be quite a distance.

Hausdorff: The nearest subway station is at the corner of 42nd and Eighth Avenue. That is only a block and a half from the Holy Cross

School, on the shortest path in the Manhattan metric from the point of sale to the school.

Justice Souter: However, the crow still has to fly down the stairs, and then down the tracks. How far is that?

Hausdorff: I'm not sure.

Justice Breyer: We can use the Pythagorean Theorem to find the total distance traveled by the crow. It would fly in a straight line from the school to the subway entrance. Then, as it flies down the stairs, assuming we know the vertical and horizontal distance traveled, we apply Pythagoras to determine the diagonal distance. Then, the crow flies straight down the tunnel along Eighth Avenue to where the train is at the time of sale. So we add on that distance.

Justice Scalia: Now honestly, what are the chances that there is only one set of stairs to get to the platform? Haven't you been to New York? Don't you know how the stations are set up?

Justice Breyer: Well, then, for each set of stairs, the Pythagorean theorem can be utilized to determine the distance. It is still a simple sum of distances that needs to be calculated.

Justice Stevens: I believe that station has only one set of stairs.

Justice Scalia: Clearly, the original intent of the legislators was "as the crow flies" or "as the mole digs," What else could they have been thinking? It would be utterly irrational to think otherwise.

Justice Kennedy: I see no reason to believe that the framers of this law even considered the question of moles versus crows.

Justice Ginsburg: For that matter, how would the crow hand the drugs off to the mole to begin with? And even if the mole did manage to dig its way down to the subway tunnel and claw its way through the subway tunnel wall, wouldn't it be immediately crushed by the speeding subway train?

Hausdorff: Yes, your honor. I believe it would.

Justice Scalia: Presupposing an inability for crows to pass drugs to moles flies directly in the face of the constitutional mandate. One can certainly imagine a trained pair of animals in the employ of a nefarious drug ring that behaved in exactly this manner. The drugs could be placed in a pouch with Velcro on the side and the bird and burrowing mammal could have Velcro harnesses to which the pouch could be easily attached and transferred.

Justice Breyer: How would the bird get the Velcro pouch off its harness at the time of transfer?

Justice Scalia: The mole could help.

Justice Breyer: Moles don't have hands. They have little flippers.

Justice Scalia: Don't you think I know that?

Chief Justice Roberts: Perhaps we could return to a discussion of the original case at hand.

Justice Scalia: Oh, the chief justice steps in.

Justice Thomas: (Snicker).

Chief Justice Roberts: I am just trying to speed this process along.

Justice Scalia: Listen, Mr. Chief Justice. I am only going to say this once. If we want to talk about moles, we talk about moles. If we want to talk about the most recent episode of American Idol, we talk about that. We're the Supreme Court, for God's sake. Just because the President picked you for the big seat ahead of all us much more experienced jurists doesn't mean you've earned the right to lead a Boy Scout troop, let alone this august group.

Justice Thomas: Hear, hear.

Justice Alito: He was just trying to point the discussion in a more fruitful direction.

Justice Breyer: I'm with Nino on this one.

Justice Scalia: Don't call me Nino.

Justice Kennedy: Personally, I am more concerned with the time element. The subway train is moving quite fast. How do we identify the instant at which the sale takes place? Is it when the money changes hands or when the drugs change hands? This could dramatically impact the distance between the school and the point of sale.

Justice Scalia: Don't you know anything about drug deals? The money and drugs change hands at the exact same instant. Neither party wants to risk the other running off with the money or the goods.

Justice Stevens: I doubt very much that running off is a concern if both parties are on a subway train.

Justice Scalia: What do you know about it? Have you ever been on a subway during a drug deal? Do you know the first thing about it?

Justice Souter: Perhaps we could take the time of sale to be the midpoint between when the drugs are exchanged and when the money is exchanged. Then if the exchange is simultaneous, that midpoint becomes the time of exchange. So everyone is happy.

Justice Kennedy: What happens if the money changes hands and the drugs never do?

Justice Breyer: Then logically, this would put the midpoint halfway between the time of money exchange and infinity.

Hausdorff: I believe that would also be infinity, your honor.

Chief Justice Roberts: I see great difficulties with legislation that allows drug sales to occur at times infinitely far in the future.

Justice Scalia: I don't.

Justice Souter: What about the velocity of the train itself? According to Einstein's theory of special relativity, length dilation will cause the distance from the train to the school to be greater as measured by an observer outside the train than as measured by an observer in the train.

Justice Scalia: Relativity does not once make an appearance in the constitution. Hence, it should not be considered here.

Justice Kennedy: It's not even an issue unless the train is traveling at speeds approaching the speed of light, which it clearly is not.

Justice Souter: But if the measurement of the distance to the school comes in extremely close to 1000 feet, then the relativistic effects can impact whether or not the sale took place within 1000 feet of the school, even at these slower velocities. So we do need to determine whether the legislators intended the distance to be measured by an observer in the frame of reference of the train or in the frame of reference of the school. Counselor, any thoughts?

Hausdorff: Um, I'm not sure what they were thinking with regard to this, your honor.

Justice Thomas: How about the Calabi-Yau metric? That's a good metric.

Justice Alito: Is no one here concerned about quantum effects?

Justice Scalia: I'm not.

Hausdorff: Perhaps, I could interject a few remarks at this point.

Chief Justice Roberts: Don't bother. We're good.

Madhur Anand

Dissecting Daisy

*Elemental analyses of the orange paint showed
that it contains particles largely made up
of sulphur and arsenic*

This is how Dutch masters rendered an orange.
Its origin
like many others
at first seems unpalatable.

When did metaphor appear in the fossil record?

New worlds invent
the common orange
lichen, *Xanthoria parietina*
long before angiosperms.
Colour precedes its name.

Foreign fruit inspired
the old world palate
and we renamed our palette.

But for things resembling
a still and studied life,
metaphor seems to fail us.

There is hope only for *sempiternal beauty*,
so Linnaeus quoted Pliny on *Bellis perennis*.

Now Daisy finds herself as we find her:
Picking the same *tiny star-shaped flowers*
of family *Asteraceae*
year after year.

Sipping orange pekoe and
waiting for a renaissance,
not just another sunburst battling night.

Notes:
1. First stanza from: Wallert, Arie (ed.) (2000) *Still Lifes: Techniques and Style: an Examination of Paintings from the Rijksmuseum*, B.V. Waanders Uitgeverji, Amsterdam.
2. "Geoluhread" is Old English for the colour orange and was used until the English-speaking world was exposed to the fruit (via France) sometime in the 16th century. The modern colour name comes from the fruit name, originally a Sanskrit word, because the fruit has origins in South East Asia.

Sandy Bonny

Frames

If you don't leave your mind open, you're not going to see them. You're not going to see them through a telescope, either. The field of view is too small. That was how it started. The two of us packing up cheese and a spiked thermos and going out to sit on a sleeping bag, dark nights, midsummer.

I said, "Night is quite another world from the day." Cooler, all the cars or most of the cars on the highway gone to sleep, and while other people sat out on their porches with bug lights and Kaiser decks, Megan and me went out on the fields, past the suburb houses and the sewage treatment plant and the half-sculpted golf course and we spread my sleeping bag out, slippery side down, felt side up, and watched the sky for alien space ships.

Megan brought a headlamp and she switched it on, an hour in, to do some reading by. I sat at the edge of its light and tried to focus my eyes on the dark heads waving around us. Horsetails, wild oats. Thistles routed out. What lived there bent in the breeze, quick and green-painted-blue by the night.

By the end of the summer it didn't bend. It waved stiff and rustled around me looking up and Megan, happy to read in the quiet.

∞

Dark nights, no moon nights, with grass hoppers chirping, we could see, away from the city, the spilled salt of the Milky Way. The tipped cup of stars falling through the satellites.

Megan said the Milky Way didn't spill, it condensed. And the stars aren't falling, because every little bit of the universe is expanding. It has been for a while, everywhere. And that is not a satellite, Stephanie, it's a plane.

I saw the black of Megan's eyes, inside the blue irises turned green by her headlamp. I heard her knowing things. I told her about a walk we took to the beach at night, me and Andrew, twelve or so years ago. After his family moved to Nanaimo and my mom agreed to pay for me to visit. Grade ten.

Andrew tried to skip a rock and lost it in the water. He threw a chunk of scrap wood in next and it splashed out away from us, sending up sparkles. Haloes of yellowy green in the water. Not reflected off the water, in the water. Glowing water.

I saw magic and didn't want to be tricked.

"There," Andrew said, pointing to the edge of the circles rippling out from the log, the edge where the glow diffused. He was pleased.

I imagined glow-in-the-dark water powder. Something that could be thrown out with a board, like the packet that turned my mom's tea to solid jelly. Stiffum's Powder that my uncle brought home from a business trip. It smelled like turpentine and the tea looked like congealed bacon fat by the time the reaction finished.

I had wanted to keep my mom from drinking it but my uncle said, no, wait, let's see what she'll do.

"Nice," I said, by the beach.

"Nice, Andrew." I wanted to turn his hands over and make him show me. But Andrew didn't like being touched, and he looked happy for once. We walked back to his mom's house and drank his dad's beer.

∞

"How long are you thinking we'll wait tonight?" Megan asked, an hour into our third trip to not see UFOs.

I told her, "I'm a very patient person," which isn't quite true.

"Check out that tree." I pointed to a tree-shaped silhouette across the field. I thought she'd like that. I thought she'd tell me what it was.

She nodded.

I had Megan by moonlight, crouched and twisting my sleeping bag into whirlpools around her feet. Megan, peering through the heads of grasses going yellow and dry, going to seed and feather. Megan excited by my tree.

"It's after something," she whispered.

After seed, in soil, and rain, I was thinking. Trees come after those. Why is Megan covering her mouth?

I screamed. At the dark swoop of leaves that plunged out of the tree and into the grass. At the squeak and screech of a mouse, or a shrew, maybe, pulled into a flapping sky.

I'd never seen it outside of a television frame and bounced, breathless, risen to a crouch beside Megan. We watched them rise.

At the veterinary surgery, our hospitalized birds eat frozen day-old chicks but here was an owl, hunting in the grass, with eyes so much better than ours he could see mice when we could barely see him.
I hadn't even noticed him in that tree.
I marveled, "He must have been watching us all this time."
"What?"
Middle-earth to Megan. 'This whole time, he could see us.'
We looked up at the sky that was empty again and Megan said she'd never heard of hawks hunting at night. I said, "Me neither." I felt like I should know the difference between a hawk and an owl, but it moved so quickly up and away.

∞

I grew up on t.v. Breakfast and after school. Mice fly there, in superhero suits. Animals talk and wear clothes. Cats and dogs chase each other, laugh, cry, and sometimes rescue injured and trapped children. Zorro's horse was cool too but, as a child, I was afraid of both Toronado and the blue plastic horse with red eyes you could ride for a quarter outside of Safeway.

"Small animals," I told my mom, who did not agree that my birthday gerbil would make a good visiting pet for her hospital's children's ward. I wanted to be a doctor like her, but I wanted to fix cats and dogs and gerbils. I dreamt about them, lined up in their beds, waiting for me to come by on rounds with a special stethoscope for hearing through fur. I'd bring them treats to eat if they felt well enough. If I had a kid they could pass the treats out, the way I did on weekends when my mom had to work. There wasn't much else for me to do but follow the candy stripers around with their cart of cartoons and mints.

Three years deep in vet med my supervisor told me I needed to pick a project we stood at least half a chance of getting funding for. When I shook Megan's hand the first time my hand had just come out of a cow's rumen. It had been gloved, and was clean, but was still incredibly warm. Megan was modeling the trajectory of methylmercury through northern ecosystems. I reached into the warm wet bodies of cows to sample partially digested thorns.

Elbow deep in a live, second-hand, fistulated cow, a city kid wonders what they're doing in vet med. Neck deep in a moonlit field with a horrible conversationalist, I worried that Megan would wonder why she'd bothered coming. Then that hawk came down and caught us both up in watching for it to come back.

∞

We brought binoculars the next time we had a free night, but the moon wasn't out. The only things we could really see were distant headlights, passing on the highway, and stars. Holes to heaven. Balls of gas. Points of light spreading apart and apart in the night. Tons of them, brighter without the moon.

I did go back to the beach, Megan. The same beach on the edge of Nanaimo but this time I was as tall as I was going to be and Andrew had lost all hope of passing my ribcage. I'm not allowed to call him little. My bitter 4'8" friend. He asked me out to visit after his dad brought home brochures about bone-lengthening surgery. You could add a few inches, Andy, he said. You could reach the light switch, the paper towel in public bathrooms.

He is small, but Andrew manages fine with both light switches and paper towel dispensers. He's not that short. Though I have seen him, drunk and rowdy, jumpstart a motion activated hand dryer. He danced under it, hands up, air ruffling his hobbit hair.

But this is before.

This is when angry skinhead Andrew told me to fuck off so I did, storming down an asphalt path to the sudden dark of the unlit ocean. He wasn't a dwarf, he shouted at me. He had normal proportions, but none of the proteins that accept growth hormones, so there was never a lot his doctor (my mom) could do to make him grow. I'd told him I thought the official definition of dwarfism was 4'10" and under.

The quiet splash and the row of ripples where the scooped side of my perfect pebble touched down on the water were imagined. I pictured a stone skipped perfectly, igniting circles of light. A chain of glowing. But I've never had an arm for skipping stones. I practiced monkey farts.

Throw a rock up high enough that it can fall straight down. A perfect 10'-high board diving stone hits the water without a splash or ripple, and only a tiny, hiccupped sound.

I wanted a flash of light too but my monkeys farted in dark water that stayed dark. I pushed a piece of driftwood into the water with my foot and it rolled away, wet and black in its shadow. I guessed maybe it was too cold.

∞

I had looked it up by then. This phenomenon. Bioluminescence. I rolled the word in my mouth and ground it beneath the heel of my sneaker. Light that comes from chemical reactions inside certain species of microscopic marine plankton that don't, apparently, grow in February off the coast of Nanaimo. Bioluminescence was very cool, but it left Andrew less tricky and magic than before.

I looked UFOs up too. I learned about freak cloud formations that are perfectly natural but don't form often enough for most people to know what they are. I looked up hawks and found out that lots of them hunt at night, or at least in twilight. That it is at night that they are in greatest danger of swooping down in front of semis on the highway and having their wings caught in metal grilles. We had a lot of those birds brought into the surgery. Most vet students perform their first euthanasia on injured wildlife. I had to look into the blind eyes of a well loved seventeen-year-old basset hound.

∞

"How about a movie, Megan, next time?" I asked.

"You said these weren't dates."

"No. No, but it's starting to look like UFOs might not exist."

Megan said of course they exist, we haven't identified every object that flies. But just because an object is unidentifiable by a couple of amateur sky watchers doesn't mean it's manned by aliens.

"Ladied by aliens?" I tried.

∞

I don't think I meant them to be dates, per se. Our first trip, at least, really was about convincing Megan that she wouldn't see things she didn't look for. But our spot at the edge of the field got marked out by a rectangular patch of flattened grass. It never quite had time to recover from my need to watch and listen. From Megan's wanting to see another bird swoop down.

Our rectangle was at the end of a road that we only drove at night. It passed a fence that closed in what used to be a landfill. It was just a space inside a fence and Megan and I never thought about it. We parked ten or twenty meters away from the end of the fence and walked out into the field and sipped from our thermos and ate our cheese, listening for sounds in the world that is awake when people are asleep. I listened for the differences between the wet black ocean and our landlocked, life full, field. I listened for Megan turning her pages.

∞

I met Megan in biochemistry class but she didn't meet me. I blend into walls and I sit in back rows. I learned her name because she asked questions, all the time, every class. I thought she was beautiful. We didn't actually meet, though, until Andrew came. He showed up at the veterinary emergency doors, exploding with big and secret news, and announced that he was taking over my couch for a couple of weeks. To visit. To calm down. To avoid his dad who did not quite approve of the job he was so excited about.

I had just pulled my arm out of Mrs. A7859's cannulated fistula when Andy popped into the lab with Megan trailing hesitantly behind him. The beautiful girl from biochem who I hadn't quite worked up talking to yet, but had made the mistake of pointing out in the quad. He recognized her at the gym. He's right charming when he wants to be. She came along for the drink that Andrew and I had apparently planned to meet up for. She said, "I have two black belts," when we offered to walk her home.

Oh lanky poet veterinarian, Andy called me. Friend to all cats but afraid of cows. It's true. I had to make them eat thistles when their neighbours were munching oats. Then I reached into their stomachs to dig them back out. My cows were big and black as horses, with heavy stomping feet, and giant wet eyeballs that followed me donning my long green gloves. I waited all summer for one of them to snap and kick me.

I should never have told Andrew. But I did, and something besides about their readily infected udder systems. I told him something about being off breasts.

"Here's the lady-killer," he greeted me with Megan right there.

"Thanks, Andy, for the credit." I said, "The thought.'"

It didn't quite make sense, but soon we were hanging out. Megan always laughing, Andrew more and more worked up about his secret job. I felt five years old—awkward, excited, afraid and not afraid of Megan looking for whoever it was Andy seemed to be making me out to be. She really did have two black belts. One in karate, one in kung fu. And she was funny on top of smart.

The week after Andy left, I got stuck for something to say and asked if she'd ever seen a UFO.

∞

You can't see a tree for the lack of a forest sometimes. Miss things that are where you don't expect them to be, sort of a thing. We didn't see eleven deer all summer, Megan and me. They were right there, behind the landfill fence that we drove past again and again on the way to our field.

Twenty-eight orphans got turned in to the university that summer. Eleven past the capacity of the petting farm and the stables of a parasite study combined. Vet Sci. ranched them out at the landfill, trusting the fence that kept the fossilizing baby diapers in to keep any predators out. There was forage, and they trucked out water and extra feed. No one worried about the slightly rusted gate latch, mounted level with a four-month-old fawn's head. Level with the beginning antlers of a gangly white-tailed deer.

The man on the phone talked about blood, all over its head, in its eyes. He imagined juvies with slingshots, knives and pellet guns. I dropped a half-digested thistle into the palm of a summer student, paged my supervisor, and rushed out to the truck with a medi-kit, lasso, pvc for splinting, and ever handy duct tape. Dr. Ed brought a gun, tranquilizer cartridges, and bullets. In case, he said. But the buck was a tiny, trembling thing with stick skinny legs. Tendons taught behind its knees. And half the blood on its head was rust from the gate latch.

I caught him in a headlock. Held his bony shoulder against my ribs while he flapped his Bambi ears, stiff as rubber, against my neck. He was stronger than I'd expected, smaller but jumpier than my cows, and I had to pin him against the fence to keep him still while Dr. Ed cleaned his head, gave him an injection of antibiotics and another for tetanus, and left the gun lying in the grass. Metal shining black as the little guy's nose. Black as his eyes, focused ahead on his friends, who had backed away into the far corner of the lot. They were tall, outgrowing their speckled baby coats and almost ready to be shipped out to a preserve away from the city. I whispered that to the fawn, "Three weeks left," before we let him go.

We capped the gate latch with pcv, so he couldn't worry his head anymore. Then we turned to watch him join the other deer. They approached him, then sprang away from the strange smell of the antiseptic, from the smell of my armpit lingering on his neck.

∞

"They're bigger than I expected," Megan said, when I shone the flashlight through the fence. They'd never had to worry about food or running away from anyone but me and Dr. Ed.

"They are bigger than wild fawns, my deerling," I said.

But Megan was making cat noises, beckoning. They pressed their soft mouths up to the fence to take grass from her hands. They scattered when I started the car, then turned to watch us drive away. Twenty-two eyes reflected our headlights, coupled coins in a linked metal net.

Megan by movie-light frowned seriously at the short that told us to turn off our cell phones. Our seats were linked and she frowned at me shifting, forwards and back again. She reached her tongue out, deer-style, and pulled it into her mouth with popcorn clinging to the tip.

I shook my head, I couldn't eat.

"Relax," she whispered, "It's a long movie."

It was a long movie, and for most of it they shrunk normal-heighted actors into hobbits with computers magic, but they hired real stunt doubles for the action scenes. Real stunt doubles to fight the Orcs with the Ents.

"S'Andrew!" I shouted, with pointing. The people around us stared.

But it was Andrew. From the front, just some actor acting. But from behind, Andrew jumping out of the arms of an Ent and landing in an awesome shoulder roll. It was Andrew. Lifted out of our life and into Tolkien's imagination. We poked each other, again and again, grinning in the dark, whispering, "It's Andy."

He's the one who imagined us first, Megan and me. Our hands brushing together in the popcorn, our fingers touching and rustling seeds.

Wendy Brandts

Collisions

[*A woman, on her knees beside a potted sapling, is digging frantically. Her gloves, briefcase, keys, are strewn over the icy driveway. She pauses, looks up.*]

∞

Hear that? This hamlet—these zigzags of snowy rooftops blue with shadow—is hesitating. And see the clouds? They are weeping crystals! How can those cars, impatient commuters, glide by so unblinking, unaware, indifferent?

I dig, and I dig for answers. But this soil is already frozen, inert as a comatose mind. Silent as my boy in the hospital.

A few degrees of frost, ten points on the TSX, a six A.M. summons into my office. Don't take your bike today, I said, leaving him at his cereal bowl this morning. Who could have predicted the impact?

It is not like the weight, calibrated on a scale, of gravity pulling to the centre of the earth: a universal force experienced by every root, stone, foot on the ground, every bird, snowflake, star in the sky. Nor is it a measured hit: a vehicle hurling a crash-test dummy at a designated velocity. No, this impact reverberates without end. Splinters of splinters of splinters shatter the infinite facets of our future. Scar all the faces of our past.

My mother, touching my shoulder and whispering, He'll be O.K. They'll find a way. Her faith.

How we are confused and deceived! Statistics. Hope. We've heard of it before: Desperate Man healed by holy water, the Poor Woman who divined the jackpot, an Innocent Child passing through fire. Survivors—like us—clinging to unlikely rafts. Proof enough.

And we know the probability of pregnancy, the rate of collisions, the chance of snow.

But individual actions—and consequences? Whether that man will use his fist again, or this mother will decide to leave before the snowstorm; whether another man, distracted by his cell-phone while he drives, will glance away as a sudden greyness engulfs the sun; whether this boy—my boy—on his bicycle, blinded by shards of ice, meeting that car on the hill, will swerve and fall.

We do not even know whether the simple actions we take day after day will lead to: sustenance, a smile, a memory. How can we bear, at each instant of our lives, to face the outrageous proliferation of possibilities? The subsequent collapse into the actual?

I plant this oak tree, but I do not know who will see it grow. Will my boy return? Will I? Who will nurture this tree? Remember it? Two, four. . . six, seven, eight. I can count the branches. Feel the weight of the roots. But how many leaves, how many years, are furled inside?

My nails gouge this hard earth. [*Clenching her hand.*] I can toss away black fistfuls of history: decayed roses, buried pets, boot heels, tea leaves—the matter remaining after form and function have departed. I can give reasons why this sapling is standing here: because of bark and soil, because of encroaching winter, because of the tree nursery, my mother, my job, this street, the radial force of gravity. . . and my child. But as it stands, thin and leafless on this hill, as this purple November evening, caught in a street of picture windows, fades into the blue flickers of local news, all we really know—all we will ever know is: today, a boy, this first snowfall, an accident.

Active Pass

Dusk whelms the ferry, peels from the sides, laminar flow husking into eddies, dusk, that slow-moving shadow shockwave. Five minutes out of Active Pass, early April; the ship ploughs East across the Georgia Strait; I look back. Ocean, acres of momentary diamonds rise and fall, resemble the almost random hum under thought. One water patch coheres: the waves swell in unison, echo light, then hollow, tilt and pitch, and in a flash I see they are not waves, but birds. Some moments incandesce. Time crests and shatters and the shards, salt dust, almost hush, embed in your heart. I settle into breath, quiet, not quite silent. Two bird clouds surge apart, mass together. Their bodies bank, mirror the last yellow-steel and vanish against the water, then swing past the angle of reflection and resurface. Their wings are crescent shadows under wave crests, except they move as one: pull, flow, contract, one muscle, mind, from many. In their swirl of appearance and disappearance, there is enough to fill a life.

The deck shudders; children gallop and shriek; off the bow, radiant points spangle the shore, the city's roar muted by twelve kilometers of air. I glimpse where I'm going, glance down into vortices, swirls punching a downwelling line laceworked with foam, which luminesces in the moment before nightfall. Not all things vanish into darkness. Some vanish into light. I watch the birds lift and tumble, a lip of mercury rolling as it rises, until I cannot tell myself from sky.

Magic Stretch Glove™: 85% acrylic, 15% spandex

You think you invented it? You're wrong. Fusion of dinosaur blood and glycerine, three steps off from nitro, it morphed up through the Precambrian, black gold into crimson knit, to show up on the shelf, ninety-nine cents at Zellers. How would it think? you think. You think inside, glove thinks no sides, thinks continuously deformable, thinks parachute for the mind. You think it wants your hand. You're wrong. It wants emptiness. If your paw weren't stuffed in it, the glove could strain secrets from the wind, could feel up the inside of nothing. It made itself in your hand's image because everything makes itself in something's image—the turtle in the image of stone; the moth, birch bark; you, nakedness.

Tyndall Field
A Catalogue of Stars

We lie in the night, in a rustle
of names: Arcturus, Canopus,
Sirius, the hiss
of starlight on skin.

Procyon, Polaris,
Pulcherrima, pull us
from ourselves

into the swell of dying giants:
Aldebaran, Bellatrix,
sighing off our shells,
this too a death

in the damp grass, naked,
naming, touching,
and naming: Antares,
Al Niyat, Alnilam, eyelid,
cheekbone, lips,
fingertips,

Sulafat, Fomalhaut,
Alpheratz, fingertips
lifting

Sulafat, Fomalhaut,
Alpheratz, fingertips
lifting every pore
into light.

Resonance
Two Experiments

(I)

It's best studied naked, preferably after making love. Lying down, press your ear to a mattress and hum. Stop after each note to listen. The steel springs catch and hold your voice; their song peaks on a certain note, then dies away as you scale higher. The note the mattress sings most strongly—maybe A flat, maybe middle C—marks its resonance frequency.

A standing wave in a kestrel's leg-bone, rigging droning in wind, resonance is the hum inside hunger, muscled space, a certain slant of sound. Everything—from cello strings and cardboard tubes to tungsten atoms—has a frequency at which it resonates, at which it hears and answers the call of the world most intensely. Light, too, sings to everything it touches: the hollow below an anklebone, river-wet basalt, electron clouds in salt crystals. Atoms and molecules soak up and spout drops of resonant light. Oxygen croons in green, nitrogen in blue, sodium in yellow. Similarly, the world sings through each of us. This is called voice.

(II)

Find a quiet place and two tuning forks. Two-pronged ears, engines of praise chanting like stainless steel monks, they can teach you how to listen with your whole body.

Mount them on separate hollow boxes and place them next to each other. Don't let them touch. If you strike one, the other will begin to sing as the pressure waves from the first reach and buffet it. This is love.

If the two forks have different natural frequencies, this energy transfer—resonance—this drawing out of voice, will not happen.

Next, stuff the hollow boxes with whatever you can find: half a dishtowel, cotton batten, your ex-lover's lost sock. When you strike one fork, its song will be muted, tamped down, so faint the other fork may not sing at all.

Now consider this: the heart as echo chamber, as tensed emptiness. The heart, too, is hollow, and so sings.

There's something I need to say to you
but I don't know what it is

A gust in the hall, my heart an empty
doorframe, a drift of laburnum petals, a shoe rack.

The swing lifts, upward sweep ending in a lurch,
momentum's gasp, the simple harmonics of want.

All I need: a rusted siltstone
in my palm; your name, your name.

Light shifts through rain. At dusk,
a stem of Mexican jasmine on a plywood table.

At night I wander through moonslicked grass.
I wear a necklace of hummingbird bones.

*

This morning I pulled a galaxy
from my chest: a string of stars, a word.

Poplar leaves flash. A tongue of wind
parts wheat. Earth shudders, opens.

Random motion: pollen grains dance in water, your hands
ghost my skin, Einstein's drunken walk.

Evidence for God: wild carrots,
a mouse's clitoris, an ant's tongue.

Fawn lilies, a rumble of white along the road bank.
Knee-deep in thunder, I wait.

*

A clatter in the hall, my heart an empty
beer can, an ironing board, a dustpan.

Your name, a possessed noun,
and me with no pronouns.

Day, being lonely, holds hands with herself,
cups darkness in her palms.

I open my mouth as if to speak.
You raise your smile, a shield.

Longing: a drawer full of spoons,
but no soup.

*

Dogwood petals float, scalloped milk gathered
into oxblood tips. Curled on my tongue, a word.

There are many kinds of nakedness. You
are almost animal in your innocence.

Sparrow prints on the dewed rail at dawn:
the hollowness of light.

All I want: to contain the sky,
that limitless blue equation. You, too?

*There is nothing in the world except
empty, curved space.* Your hands.

*

The rake rusts, voiceless, against the shed.
Even loss has geometry.

On the tray where my bracelets were:
a bowl of beach pebbles.

Water velvets the purple petunias.
Summer's last leaves uncurl like light.

What I would like:
to want nothing from you.

Something falls. A coat hook naked.
The self, a rain shell crumpled on the floor.

Praying to the God of Leavetakings

You who stand outside the flood of time and landscapes' fractal hills
to witness what escapes, mathematician of leavetakings that endlessly
repeat, decay, what prayer should we make? You outside the multiverse
with eyes to see every fallen thing, each sparrow dropping in the talons
of the sparrowhawk, each branched old-growth tree emptying the canopy,
each handful of red rock the river bears away, each of us in the grip
of great old age or sudden accident, each galaxy spiraling into the dark,
central, hungry black; you who see all forms, stones, stars falling away,
departures in kaleidoscope display—teach us to invert your sight,
to see, in our small span, the bright, brief presence of the living.

Brave New Biosphere

The pieces are in our hands—the four-letter code,
the genomic list—here's you, here's me,
thirty-five thousand genes on the microchip.
Why not augment our lives? Add the prey-stunning
electrical cells of the eel to the human arm, bioluminescent
eyes for the dark from the ocean depths,
plant chloroplasts that allow our skin, green now, to feed
on light. No need for food or oxygen;
we could stand about in the future, ferny-fingered by day,
flashing our glowworm codes by night,
needing only filamentous, permeable feet to bring water,
a waxy skin to shed wind and mosquitoes,
a little space in the fattening light. Tended, if we were lucky,
by gene-altered trees grown muscular and agile
on actomyosin, shuffling along on knuckled roots, responding
to our chemical signals of blossoming or distress
in the greenhouse gas; and if the changing shape of our future
makes us as silent as bristlecone pines, oh leave us
those cheerful and modest descendants of dinosaurs, the birds.

The All of It

The universe is modeled there
on the NASA computer screen
against a dark blue backdrop of ignorance—
all we know of space,
all we've had of time:
a drinking glass with a flaring rim
on a wobbly base, that blip
where we began inflating
from sub-atomic scale to grapefruit size
in the first trillionth trillionth of a second;
a glass of time, holding a small level
of espresso dark, and then the early stars,
fusion furnaces turning on,
and later yet the Milky Way,
and rotating about a minor sun,
a rock with water on it
where now we peer about
in the quantum foam—
and then the widening lip
where one day time will spill us out
and no lip of ours can drink again
the knowledge of that spring.
Still, the dark blue backdrop
offers hope of god or natural law
where beginnings are small enough
to hold us all, the way the mind
can hold the drinking glass
or the newborn child, that love
set going from incomplete halves.

Guided

I remember Norberto
hadn't brought his white cane
so going out for coffee
    on unfamiliar streets
he gladly held for guidance
    my gladly offered elbow.
And in mathematics, it was the blind leading the blind!
Whenever one of us had guidance to give the other,
it was gladness to be giver,
    it was gladness to be receiver.
"I see, I see," Norberto murmured.

Our Study

If you had needed to be told, I was willing to tell you
but what might I have said?
First, welcome (but I did say that). Then?
Welcome to study, and our study
is wonder and also is doggedness.
We are as if students of plants who could also be the plants;
astronomers but also planets.

It is academic,
but is our discourse timeless? If I had tried to say so,
you would have known better. You could see
    sometimes we race, after all,
and when we finish a job we may put a date on it
as if we were bottling milk.
What is a theorem's shelf life?
Some of what was real to me once has now become
not error, quite, but less real. Also,
what we know is true
would no doubt have been true ten years ago
    if we had known it then.
But two hundred years ago? Some theorems
have a shelf life in the past direction too.

Presence

Please don't make fun of me
if you should find me talking to her
as if I thought she was listening.
All very well for you to go credulous, is it,
but I should know better? No doubt.
I know she doesn't answer prayers. Obviously not.
The gambler's "Be there, baby" over his dice-roll
    is fruitless.
To her overwhelming bounty if my paean rises, it is
    unwilled and unthought.
Stranger is to find myself
everyday chatting with her.

Do I love her? Not as I love a person,
even my mate, not as I love myself.
But as I might love a most trusted horse
    embrace her smooth shoulder
    relish her warm breath
or the belovèd mountain slope
    which bears my sleeping-bag on her breast
    steady with a world's sureness.
Surely though I command my horse
    her power and step are beyond my directing
and surely though I will find my trail on the mountain
    the mountain meets her own schedule
    on times I do not reckon.
So with her whinny in our ears, saddle up
    and on our way,
vast though she be and finer-grained and longer-spanned;
be she steed, or world, or wraith.

Cold Comfort

If I took alarm at the prospect
of things spinning out of control
 (and I might
   for they are
     oh, I well might)
this refuge would tempt me.

To fall back on the cyclicity:
Orbits are periodic, things come round,
the mathematics tells truth about these worlds
and I would take comfort, only to find
that a perfectly Newtonian solar system
   with fixed unvarying laws
need not repeat, can steer deterministically
to disasters old or unprecedented.

To fall back on the optimizing tendency:
Natural selection has left and will leave in place
   just that which works,
the modern synthesis tells truth about evolution
and I would take comfort, only to find
that a perfectly Darwinian-Mendelian biota
need not direct itself toward any particular adaptation
toward any pre-defined perfection
with any particular probability
     (could any statistical conclusion be sure?
     yes, but not this one).

To fall back on predictability:
All is caused, nothing will be forthcoming

but what is embryonically already here,
the mathematics tells truth about emergence
and I would take comfort, only to find
that a perfectly Laplacian automatic universe
as we labour to know it,
as we fill in past data neatly and diligently
fills in the future (though exactly) only slapdash,
with holes you could drive a hurricane through.

The mathematics tells truth about the world.
We are its ventriloquist, yet
     some words it won't let us put in its mouth.

If I took alarm
(and I might
    I do)
this refuge would tempt me anyway.

To fall back on self-will.
While not answering all questions
to have our choice of which questions to ask
     including some big ones
     not all
and sit in our refuge spinning our little tales,
     our truth.

The Birth of Celestial Mechanics

Nature and Nature's laws, lay hid in Night:
God said, Let Newton be! and it was light.

Alexander Pope

Christmas day 1642, at dawn.[1] A grey limestone manor house in Woolsthorpe, Lincolnshire, where a woman is in labour. Within hours, she delivers a boy.

He was so little that they could put him into a quart pot, and he was so weak that they wrapped a bolster around his neck to keep his head on his shoulders. The midwives lost all hope: for a week the baby's life hung in the balance. But he made it. On January 1, 1643, the child was baptized. They called him Isaac, after his late father.

The boy didn't have much of his mother either. When Isaac was three, Hannah Newton married an old bachelor, the reverend Barnabas Smith, who convinced her to leave the child with his grandmother and move alone to his manor, three miles away from her own. The kid grew lonely and withdrawn, deprived of warmth and love. In 1653, Barnabas Smith died. Hannah returned to her old house along with two girls and a boy, the children of her second marriage. Was this a happy event for Isaac? Even if so, it proved short. Less than two years later, he was sent off to Grantham grammar school, seven miles north of home.

In the beginning, his intellectual brilliance worked against him. Physically weak, he avoided the rough games his schoolmates used to play. The boys ignored him first, later they became hostile; therefore Isaac felt better in the company of girls. As he grew older, his friendship with Miss Storer—a younger schoolmate—developed into deeper feelings. But this

---

[1] Since England used the Julian calendar and had not yet adopted the Gregorian calendar, as most of Europe had, the English dates before 1700 were 10 days behind the ones on the continent (after 1700 the delay was 11 days). So 25 December 1642 would have been 4 January 1643 in the rest of Europe. Also, until 1752, the new year started in England on 25 March.

seems to have been the first and last romance of his life. The disappointment he often felt in relationship with people made him sober, silent, thoughtful, and absentminded. Smiling seldom, having little sense of humor, he began to develop a complicated, reserved, and neurotic personality, which would frequently push him to the verge of nervous breakdowns.

As a gifted but unhappy child, Isaac found refuge in work. Skillful, intelligent, imaginative, he enjoyed building models of wood and liked finding new games. He made a mill, sundials, work boxes, toys for the girls, even a kite with lanterns to scare the villagers at night. When he grew older, he filled his time studying mathematics and philosophy. At school he quickly became fluent in Latin—the language he would later use to write his notes, papers, and books. [2]

The political and social life of England during Newton's childhood and adolescence was marked by internal instability, power struggles, and religious change. The attempt of Charles I to suppress Parliament started a civil war in 1642, which ended seven years later with the king's beheading. The English Republic, the Parliament's dissolution, the dictatorship of Oliver Cromwell, and for a short time of his older son, Richard, were followed in 1660 by the monarchy's restoration, when the House of Stuart returned to power through Charles II. Though far away from the political and military battlefield, Newton grew up surrounded by turbulent events. [3]

In June 1661, he was admitted as a student at Trinity College in Cambridge. In spite of his mother's dowry and substantial inheritance, Isaac had to earn his living waiting on richer students. This was the second time Hannah Newton-Smith refused to finance her son's schooling; as an uneducated woman who could barely write, in 1659 she had withdrawn Isaac from Grantham to avoid paying the school taxes, pretending to need the teenager's help in managing the family estate. She sent him back only after the school's headmaster together with her brother William, a cleric and Cambridge alumnus, repeatedly insisted that Isaac should return to Grantham to prepare for the university. [4]

In spite of financial difficulties and two years during which he kept a low profile, Newton soon became one of the best students at Trinity.

---

[2] Westfall, Chapter 2.
[3] Churchill, Vol. 2.
[4] Gjertsen.

Again, his brilliance hindered his attempts to find company. Though only one year older than his classmates, he was left outside their circle. Except for Wickins, his roommate, nobody socialized with him. From time to time, when in need of human contact, he played cards in a nearby pub. This solitude gave him time to concentrate on his studies, to read, and to meditate. The curriculum at Cambridge included Latin, Greek, and Aristotle: his physics, logic, ethics, and philosophy. Though not studied in class, Descartes was much in fashion, and Newton got soon acquainted with the revolutionary ideas of the French philosopher. He also read some of Galileo's work.

Newton used to take notes and analyze what he had learned and understood from his reading; he posed problems and tried to solve them. Some of the questions became significant for his scientific career: he would pursue them his entire life. He was twenty-one when he entered in his notebook the title *Quaestiones quaedam philosophicae* (Certain Philosophical Questions). These notes marked the beginning of his long and fruitful research, which would establish him as the most famous scientist of all time.[5]

1 The Notebook

Commenced most likely at the beginning of 1664 and finished late in 1665, the Philosophical Notebook is important for understanding the early thought of Newton. With respect to its content, the Notebook can be divided in two distinct parts. The first thirty and the last forty pages consist of mere notes in Latin and English on Newton's readings of—during his time—well-respected authors like Aristotle, Eustachius, Descartes, Magirus, Vossius, and Stahl. The central forty-eight pages are the kernel of the Notebook and contain the *Quaestiones quaedam philosophicae*—a first attempt to comprehend notions like *attraction, gravity, motion, fire,* and *light*. From the way these notes were written we see that the initial thirty-seven entries developed eventually into seventy-three. Some of them remained nothing but headings: Newton did not elaborate on notions like *fluidity, stability,* or *humidity*; the comments on others, like colour and motion, cover several pages. The quality of his notes also varies: some are nov-

---

[5] Westfall, Chapter 3, and Gjertsen.

el and deep, others—simple and naive. This early writing is a mixture of genius and immaturity.

Under the headlines "Of Motion" and "Of violent Motion," Newton recorded for the first time his thoughts about what would soon become the science of mechanics. For example, he remarked about *motion*:

> Descartes defines motion in the second part of the *Principia Philosophiae* to be the translation of one part of matter or one body from the vicinity of those bodies which immediately touch it and seem to rest, to the vicinity of others.

Later on, in the Introductory Scholium to Book I of his *Principia*, he came up with the concept of an absolute reference system, formed by three fixed stars. He needed this idea to provide his own definition of motion. In Newton's opinion, motion made sense only with respect to an absolute system. Indeed, if while sitting in a train in some station we see the neighboring cars moving slowly on a parallel track, which train is actually moving? To answer this we look at the platform. So motion must pertain to a reference system—the platform. But the platform moves with the earth and Newton wanted an absolute frame, so he chose three fixed stars.[6] During his time it was not known that stars are moving with respect to the center of the galaxy. Although the philosophical implications of an absolute reference system were already questioned by Leibniz, Einstein would later show that one way to avoid the difficulties is to replace classical mechanics with special relativity.

Endowed with a sharp critical spirit, Newton attacked Aristotle's arguments concerning the shape of a projectile's trajectory, explaining that the motion is influenced by what he called *natural gravity*. Newton's understanding of this concept in the Notebook was different from the notion he would later define in *Principia*. As for the Greek philosophers, *gravity* meant for him weight or heaviness. The early cosmological models of Aristotle and Ptolemy divided matter into four classes (or *elements*) ordered from heavy to light: earth, water, air, and fire. The natural place of the elements was the center of our planet, also considered the center of the universe. Fire, the lightest among them, had negative gravity, called *levity*, which suggested a tendency to move away from the earth's center.[7]

---

[6] Wightman, p. 115.
[7] Westfall, p. 94, and Collier's, Vol. 11.

A statement that illustrates Newton's physical intuition appears somewhere else in the Notebook, where he claims that the cause of gravity is heaviness, which makes bodies fall by acting on their inner substance, not only on their surface. Newton felt the need of some mathematical techniques to investigate his claims. This was probably the time when the idea of a proper mathematical theory—which would later develop into *differential and integral calculus*—found a place in his mind. He understood that mathematics was a language for the laws of nature, the key that would open new perspectives for him, so he focused on it during the following years.[8]

In 1663 Newton fell upon Euclid's *Elements*, which he needed in his study of astronomy. The first theorems he encountered seemed obvious to him, so he deemed the book a trifle. But he changed his opinion soon. After working through some more difficult results, like Pythagoras's theorem, Newton began to appreciate Euclid. Then he began to study Descartes's *Geometry*, which challenged him more. He read a page or two, until he failed to understand. Then he went back, read those pages again and prevailed over the first difficulties. But a few pages later, another statement would elude him, so he would reconsider everything from the start. Alone, without help, with undefeated energy, with will and power, he overcame all the impediments towards mastering Descartes's ideas. The geometric language and way of thinking, supported by the algebraic rules of computation, would become from then on Newton's ally in mathematics, optics, and mechanics. With an amazing geometric insight, he would soon create the differential and integral calculus, which would open the gates of modern science.[9]

Though Newton was undoubtedly self-taught in mathematics, at least one person had some direct influence on him during his student years. In 1663, Trinity College established the Lucasian chair of mathematics, and its first professor, Isaac Barrow, delivered his first course in March 1664. Barrow was not Newton's tutor, but Newton attended some of Barrow's lectures. It is likely that Barrow encouraged Newton to read the work of John Wallis, a contemporary Oxford professor, who had written in 1655 a significant book, *Arithmetica infinitorum* (The Arithmetic of Infinity).

---

[8] Westfall, p. 94.
[9] Westfall, pp. 98-99, and Pemberton, Preface.

Beside the works of Euclid, Descartes, and Wallis, Newton also commented in his Notebook on the *Geometry* and the *Miscellanies* of the late Frans van Schooten, a former Dutch professor in Leyden, as well as on the algebraic work of the earlier French mathematician François Viète. From all these extracurricular readings Newton gained a solid background in the important mathematics of his time. [10]

In spite of his erudition, Newton had a hard time obtaining financial support to continue his studies at Trinity. His first attempts in this direction failed. But he secured a fellowship in April 1664, apparently due to Isaac Barrow's involvement in the award committee's decision. This award ended Newton's pecuniary problems. From then on he could dedicate all his time to his work. [11]

During 1664 and the beginning of 1665 Newton solved several difficult and important mathematical questions he himself had raised. His progress in research proved fast and spectacular. But his first great results came after the summer of 1665, when he took refuge in isolation at his manor house in Woolsthorpe. Cambridge had been hit by plague. [12]

2 The Plague

Wars, floods, fires, and earthquakes have ravaged humankind throughout the centuries and continue to haunt us in modern times. But in spite of the evil they bring, these disasters also unite us in our effort to overcome them. No catastrophe, however, seems to have been more cruel and more devastating than plague. Its psychological consequences were often more damaging than its actual power of destruction. An infectious terminal disease without cure divides us, isolates us, and it shows what easy prey we are if alone, unprotected by the strength of society.

The earliest recorded epidemic, which appears in Chapters 5 and 6 of the Bible's Book of Samuel, hit Philistia, in ancient south-west Palestine; another one swept over Lybia, Egypt, and Syria around 100 A.D. The first epidemic of high proportions devoured Europe in 542 A.D., during the time of the Emperor Justinian; the second pandemic, known as the *black death*, spread over Europe in the fourteenth century, reaching its height between 1347 and 1350; the third and last big wave ravaged Asia

---

[10] Westfall, pp. 99-100, Gjertsen, pp. 604-605, and Feingold.
[11] Westfall, pp. 101-102.
12 Westfall, pp. 105-140.

and extended to parts of the Americas at the turn of the 20th century. The so-called *Great Plague* of 1665, which seized Cambridge, was limited to England—mainly to London and its vicinity.[13]

But since no event is only evil as no event is only good, the London epidemic had a catalytic influence on Newton's work. Once the plague struck, academic life in Cambridge fell apart. Joined by their tutors, many students continued to take lessons in neighboring villages; others went home—Newton among them. In March 1666, when the first phase of the epidemic ended, Newton returned to his college, but in June, when the plague hit anew, he left for Woolsthorpe, to rejoin Trinity only in April 1667. During this period of seclusion, he found time and peace of mind to concentrate on mathematics and start building on the results that would later make him known all over the world. These eighteen months were among the most successful and prolific in his entire career, shaping his future work in mathematics, mechanics, philosophy, and religion. Historians of science refer to this time as the *annus mirabilis* (the wonder year), 1666.[14]

In fact, although the plague and the isolation at Woolsthorpe helped Newton lay the foundation for his achievements, the Notebook and the papers he wrote in Cambridge in between the two phases of the epidemic show that the isolation created by the pestilence, though helpful, was not crucial in focusing his mind. His intuition, energy, and concentration power, his enthusiasm and ambition, would have probably helped him succeed anyway.

But very few people knew about his results, for Newton kept them secret. His plan was to develop them, let them mature, and then present them as a finished product later on. This was probably out of fear that someone else might use his ideas and publish better results before him. The dictum "publish or perish" had not yet been invented.[15]

During the time of the plague Newton focused on research. The farm was self-sufficient and provided him with a comfortable life, so he did not worry about tomorrow. His main preoccupation was his mathematics. He could pursue it without any equipment, test, or physical experiments, he enjoyed it, and he knew that it might help him develop his ideas regard-

---

[13] Westfall, pp. 105-140.
[14] Collier's: Vol. 19.
[15] Westfall, pp.143-144.

ing some of the questions in the Notebook. Among the new mathematical results he came up with in these months of seclusion were the binomial formula and the tangent method for finding approximate solutions of an algebraic equation. The most important mathematical achievement, however, was that of laying the foundations of infinitesimal calculus.

Besides mathematics he also dedicated time to the issue of gravitation. The wonder year is connected with the falling-apple anecdote, told by Voltaire in the 18th century. We will never know if the idea that gravitation acts not only on objects close to the earth but also on the moon and the other planets, came to Newton while watching an apple falling down from a tree. There are many other things we will never know about him. But we can understand his accomplishments and follow the adventure of thought that took place in his mind during the *anni mirabiles* that followed.

3 Predecessors

Newton was well acquainted with the work of his forerunners. A memorable recognition of those pioneers occurs in a famous paragraph written toward the end of his life, which states that he had reached higher than others because he stood "on the shoulders of giants." Newton's mathematical, physical, and philosophical knowledge went beyond Euclid, Wallis, Aristotle, and Descartes. He knew about the achievements of Copernicus, Galileo, Brahe, and Kepler, who had broken with ancient astronomy, thus laying the foundations for Newton's advance in the theory of gravitation. For a better understanding of his contributions in this direction, let us briefly mention the conclusions obtained by these four predecessors.

Nicolas Copernicus, in Poland, had shaken the dogmatic conception of science imposed by the strong Catholic church. He explained that Ptolemy's geocentric system (geo = earth), which considered the earth to be the center of the universe, is a complicated, artificial way to describe the motion of the planets. He came up with the more natural *heliocentric system* (helios = sun), which places the sun at the center. Copernicus's theory claimed that the planets move around the sun in circular orbits, a model that offered a good theoretical approximation of planetary motion when compared with the astronomical observations available in the sixteenth century.

Galileo Galilei, in Italy, had investigated the motion of bodies under the influence of gravity, correcting several mistakes of Aristotle. Galileo's experiments showed that the time required by a complete oscillation of a pendulum does not depend on the amplitude of the oscillation; he also explained that, in the absence of air resistance, heavy and light bodies dropped from the same height would hit the ground simultaneously. In physical terms we would say today that the earth has a constant *gravitational acceleration*, a property used by Newton to formulate his gravitational theory.

The Danish astronomer Tycho Brahe believed in a model that was a hybrid of Ptolemy and Copernicus. He thought that the best way to explain the motions in the solar system would be to say that some planets moved around the earth and others around the sun. To prove his claim, he made extensive observations, more accurate than anything that had been done before. Though his model was short lived, his observations proved useful for one of his assistants, the German astronomer Johannes Kepler.

Towards the end of the 16th century, Kepler used Brahe's recorded observations of the planet Mars and compared them with the computations obtained from the heliocentric model. To his surprise he found a deviation of a few arc minutes from the circular orbit of Copernicus. The human eye can distinguish only between celestial objects that are at least one arc minute apart, closer objects appearing as a single entity. Since observations before Kepler were made without telescopes, the error that stirred his curiosity was very small for that time. But it proved enough to make him doubt the validity of circular orbits and fuel him with so much interest that he dedicated the next twenty years of his life to the understanding of this problem. In 1609 he published *Astronomia Nova*, in which he announced his first two laws of planetary motion; ten years later he brought out *De Harmonice Mundi*, which contained the third law. In these he described the basic dynamical properties obeyed by the planets of the solar system, as he heuristically obtained them from the available observations.

Kepler's laws can be stated as follows:

1. The *law of motion*: every planet moves on an ellipse having the sun at one of its foci.

2. The *law of areas*: the ray connecting the sun and the planet, sweeps equal areas inside the ellipse in equal intervals of time.

3. The *harmonic law*: the squares of the orbital periods of any two planets are to each other as the cubes of their mean distances from the sun.

The first law says that the orbit of a planet is an ellipse and the sun is at one of the foci. On the orbit, the closest point to the sun is called *perihelion* and the farthest one, *aphelion*. Most planets move on almost circular orbits—this explains why Copernicus thought that the orbits were circles and why the error Kepler found was so small.

The second law states that the segment connecting the sun and the planet (whose length varies with time) sweeps equal areas in equal time intervals. This implies that planets change speed: when far from the sun they move more slowly, when close, they move more quickly; the velocity attains its minimum at the aphelion and its maximum at the perihelion.

The third law is a mathematical relation between the mean (average) distance d from the planet to the sun and the period T of revolution, i.e., the time the planet needs for one complete trip around the sun. This law states that the quantity T^2/d^3 is the same for all the planets of the solar system.

In spite of their importance, Kepler's discoveries remained in relative obscurity until Newton understood their significance. But Kepler had known their value. In *Astronomia Nova* he wrote:

> Eighteen months ago, the first dawn rose for me; three months ago, the bright day; and a few days ago, the full sun of a most wonderful vision; now nothing can keep me back... Well then, the die is cast. I am writing this book for my contemporaries or—what does it matter?—for posterity. Has not God himself waited 6,000 years for someone to contemplate his work with understanding?

Not only did Newton grasp the importance of Kepler's laws, but he went many steps beyond them. For this he assembled the pieces built by his predecessors and added some of his own. Let us now see what was going on in his mind and how he found his way to the infinitesimal calculus and the gravitational model.

4 Infinitesimals

Many freshmen encounter difficulties with calculus. The sudden change of the objects of reasoning, from determined quantities to infinitely small ones, is a shock for most of them. This is no wonder. To grasp the principles of mathematical analysis, one has to invest time and energy. Mathematicians needed almost three centuries to put the ideas of differential and integral calculus on a sound basis. Although Newton and Leibniz, independently of one another, opened this path in the 17th century (using the mathematical work of Johannes Kepler, Francesco Cavalieri, Pierre de Fermat, John Wallis, and Isaac Barrow), the rigorous logical foundations of this branch of mathematics were not developed until the 19th century by a French and a German, Augustin Cauchy and Karl Weierstrass. And their ideas found a sound basis only in the 1960s through the work of the American mathematician Abraham Robinson. Newton's formal and conceptual use of calculus was far from what we use today.

In spite of being almost entirely absorbed by mathematics during the plague years, Newton had in mind an essentially physical goal, in contrast to the German mathematician and philosopher Gottfried Wilhelm Leibniz, who invented calculus while dealing with a geometrical question. As the Philosophical Notebook shows, Newton invested a lot of effort to understand *motion*. He related two other concepts to it: *force* and *velocity*. The former led him to formulate the three fundamental laws of mechanics; the latter opened the way to differential and integral calculus.

Newton invented the notions of *derivative* and *integral* while trying to find a mathematical way to express velocity and work. Leibniz proposed the same concepts when dealing with the tangent to a curve and the area of a region bounded by a curve. Many contemporary calculus textbooks introduce the derivative and the integral as Leibniz did, also using his notation; velocity and work appear as applications.

Newton imagined a curve as traced by a point "flowing" in time. Therefore for the derivative, usually denoted today by dx/dt, which Newton represented as a dotted x, he reserved the name fluxion, and meant the ratio between the momentum dx (i.e., the "infinitely short" path, dx, traced by the point in its motion) and the "infinitely short" interval of time, dt. He clearly lacked the approach to a limit that defines the derivative today, which was first considered only in 1823 by Cauchy.[16]

Though using the more appealing notations dx/dt and $\int$ which the scientific world later adopted, Leibniz had no better understanding of the

---

[16] Bell 1992, p. 151, and Smith, p. 614, 635.

basic notions of calculus either, compared to present-day standards. For example, to compute the differential $d(xy)$ of the product between x and y, he multiplied the quantities $x+dx$ and $y+dy$, then—without justification—subtracted xy and finally neglected the product $dxdy$, which he considered too small. Thus he obtained the correct formula, $d(xy)=xdx+ydy$.[17]

Commenting on the lack of rigor at the basis of Newton's and Leibniz's calculus, E.T. Bell, a leading mathematician in the first half of the 20th century and a former president of the Mathematical Association of America, wrote in 1940:

> History shows that frequently the essential, usable part of a mathematical doctrine is grasped intuitively long before any rational basis is provided for the doctrine itself. The creative mathematicians between Newton and Cauchy obtained mostly correct results . . . because, in spite of their ineffectual attempts to be logically rigorous, they had instinctively apprehended the self-consistent part of their mathematics.[18]

The most important contribution to calculus, which both Newton and Leibniz made, was that of connecting the notions of *derivative* and *integral* through what is known today as the *fundamental theorem of calculus* or the *Leibniz-Newton formula*. Newton expressed this result without using *functions*, which were only later introduced by Leibniz. Roughly speaking, the Leibniz-Newton formula shows that derivation and integration are operations inverse to each other: if a function is integrated and then differentiated we recover the initial function. This result was later generalized by Gauss, Green, Ostrogradski, Stokes, and others.

Newton submitted his results to the London Royal Society and to Cambridge University Press in 1669 and 1671, but his papers were rejected. Luckier, Leibniz published *Nova Methodis* in 1684, a work in which he gave the rules of differentiation for the sum, product, and quotient of two functions; two years later he brought out *De Geometria*, important for proving the fundamental theorem of calculus. The rejection of Newton's papers and the editorial success of Leibniz

---

[17] Bell 1992, p. 153, and Child.
[18] Bell 1992, p. 153.

would later lead to a priority dispute over the invention of infinitesimal calculus.[19]

But as often happens with profound ideas that emerge ahead of time, differential and integral calculus encountered strong opposition in some scientific and philosophic circles. Besides the dispute between its creators, which was taken over by their disciples, resistance to the theory itself continued even after the death of its creators. This was mainly due to the lack of foundational rigor. A notable opponent was the philosopher George Berkeley, Bishop of Cloyne, known for his idealist philosophy, based on the principle *esse est percepi* (to be is to be perceived). It is hard to "perceive" infinitely small quantities, so Berkeley's objection to calculus is no surprise. In 1734 he attacked the promoters of infinitesimals in his *Analyst—A discourse addressed to an infidel mathematician* (most probably to Edmund Halley, a disciple and friend of Newton, who had become famous for computing the orbit of Halley's comet). Berkeley found the idea of "exceedingly small" ridiculous. He claimed that a quantity is either zero or not, and in the latter case it cannot be as small as you want. Berkeley was hard to refute. Here is a sample of his argument:

> And what are these same evanescent increments? They are neither finite quantities, nor quantities infinitely small, nor yet nothing. May we not call them the ghosts of departed quantities?[20]

In fact, a century later, even the German mathematician Karl Friedrich Gauss, one of the greatest mathematicians ever, warned against thinking of the infinitesimals as real quantities. In a letter written in 1831, he stated that "infinity is only a way of speaking." No wonder that Abraham Robinson's creation of *nonstandard analysis* in 1960 was deemed by the famous Austrian mathematician Kurt Gödel to be "as important as that of the non-Euclidean geometries."[21]

A notion entirely based on calculus, whose importance Newton understood from the very beginning, and which he extensively developed in *Principia*, was that of a *differential equation*. To explain it, we will use a metaphor. Imagine the surface of a river and think of it as generated

---

[19] Stillwell, pp. 109-110.
[20] Hurd and Loeb, p. ix.
[21] Robinson.

by water particles. Each particle describes a smooth curve and all these curves form the surface of the river. Suppose that we do not know the trajectories but we do know the *velocities* of the particles at every time. From the geometrical point of view, the velocities are tangent to the unknown curves.

Under the above assumptions, a *differential equation* is a mathematical relation that describes the values of the velocities at every point of the river's surface. To *solve* a differential equation means to find the smooth curves whose tangents are known. Often we would like to know all possible curves, in which case we obtain the *general solution*. Some other times we are content with only one specific curve; then we seek a *particular solution* of an *initial value problem* (since the differential equation is accompanied by an *initial condition*).

Newton was not the first to consider differential equations. The Scottish mathematician John Napier, known as the inventor of logarithms, had solved some simple ones before. Nevertheless, Newton was the first to understand their crucial importance. As peculiar as it may seem today, he did not rush to publish his result. On the contrary, he kept it as secret as a newly found gold mine that awaits further exploration. But in order to claim priority in case someone else had the same idea, he published the discovery as an anagram, which if untangled stated: *Data aequatione quotcumque fluentes quantitae involvente fluctiones invenire et vice versa* (i.e.: Given equations involving how many soever flowing quantities, the flow can be determined, and conversely). In terms of our metaphor, Newton's statement is a vague description of the connection between the velocities of water particles and the solution curves.

Differential equations have a wide range of applications in various fields of human activity such as anthropology, astronomy, population biology, brewing, business, chemistry, cooking, cosmology, ecology, economics, electronics, engineering, epidemiology, finance, mechanics, medicine, meteorology, oceanography, pharmaceuticals, physics, politics, psychology, space science, and sports, to mention only a few. But in the seventeenth century, Newton created them with a single goal in mind: to understand the motion of celestial bodies and especially that of the moon.

5 The Gravitational Model

Before Newton, gravity meant weight. It was viewed as a local property. Copernicus and Kepler had shared the idea expressed in antiquity that each planet has its own gravity: an object close to the surface of the earth is attracted by the earth, whereas one close to Mars is attracted only by Mars. They did not think that gravitation was global and therefore all celestial objects attract each other. Moreover, they found it hard to accept that gravity is a force. Thus, Pierre Gassendi, a leading French mechanician, proposed in 1641 a model in which bodies were connected to the earth by very thin strings, the more strings attached, the heavier the body. Kepler, on the other hand, believed that his laws of motion were determined by the sun's magnetic attraction.

Descartes was the first to think that gravity extends to the moon. However, since he denied the existence of void and claimed the presence of a medium (called ether) that fills all space, he attributed gravity to a vortex, which, rotating with the earth, reached the moon. The Dutch scientist Christian Huygens shared this idea but disagreed with Descartes about the details. [22]

Though Newton was the first to think that gravitation is a force and—combining this with what he knew from Descartes—that it extends to the moon, some of his contemporaries had similar ideas. In the *annus mirabilis* 1666, another English scientist, Robert Hooke, mostly known today for the law of proportionality between the elastic deformations of a body and the forces exerted on it, tried to explain why planets move around the sun. He argued that:

> Circular motion is compounded of an endeavour by a direct motion by the tangent and of another endeavour tending to the center. [23]

And in the same year, 1666, the Italian scholar Borelli expressed the same intuitive idea in even better terms:

> Let us suppose that a planet tends toward the sun and at the same time, in its circular motion, recedes from this central body in the middle of the circle. If these opposing forces are equal they will balance, and the planets will continue to revolve around the sun. [24]

---

[22] Gjertsen, pp. 236-241.
[23] Koyré, pp. 181-182.
[24] Ryabov, p. 33.

In 1670 Hooke got even closer to the truth. He claimed that:

> Not only the sun and moon have an influence on the body and motion of the earth, and the earth upon them, but Mercury, Venus, Mars, Jupiter, and Saturn also... have a considerable influence upon its motion, as... the corresponding attractive power of the earth has a considerable influence upon every one of their motions...[25]

The scientific world was ripe for a crucial discovery. The idea that gravitation is universal had arisen.

But in 1666 Newton had already gone further than Hooke and Borelli. A paragraph written by John Conduitt, Newton's nephew-in-law, which might have suggested the anecdote of the falling apple to Voltaire, gives us a hint of what had been in Newton's mind:

> In the year 1666 he [Newton] retired again from Cambridge ... to his mother in Lincolnshire and whilst he was musing in a garden it came into his thought that the power of gravity (which brought an apple from the tree to the ground) was not limited to a certain distance from the earth but that this power must extend much farther than was usually thought. Why not as high as the moon, said he to himself, and if so that must influence her motion and perhaps retain her in her orbit...[26]

Since in Newton's vision the moon moved on a circle around the earth, he estimated that gravity might be a force that varied inversely as the square of the distance (this means that if the distance increases in the sequence 1,2,3,4,5,..., the force decreases in the sequence 1, 1/4, 1/9, 1/16, 1/25, ...). He made the computations, but unfortunately lacked a precise value for the length of the meridian corresponding to one degree of arc, which he estimated at 60 miles, so his result disagreed somewhat with the moon's recorded observations. Three years later the French astronomer Jean Picard obtained relatively accurate measurements for the meridian passing through Paris, good enough to make Newton's theory fit the astronomical data.[27] But Newton would be unaware of this fact for a long

---

[25] Koyré, pp. 181-182.
[26] Newton 130.4, p. 10-12.
[27] Petit Larousse.

time. After this failed attempt he lost interest in gravity. His attention was caught by other problems.

About fifty years later he wrote about these first endeavors:

> And the same year [1666] I began to think of gravity extending to the orbit of the moon and (having found out how to estimate the force with which a globe revolving within a sphere presses the surface of the sphere) from Kepler's rule of the periodical times of the planets [Kepler's third law]..., I deduced that the forces which keep the planets in their orbits must be reciprocally as the squares of their distances from the centers about which they revolve, and thereby compared the force requisite to keep the moon in her orbit with the force of gravity at the surface of the earth, and found them answer pretty nearly.[28]

Newton's interest in gravity was suddenly revived by a letter from Robert Hooke, dated 24 November 1679, and by another one written in January 1680. The first letter also informed him about Picard's measurements of the meridian. In the meantime Hooke had come himself to the idea that the force of attraction between planets decreases as the square of the distance, but failed to derive the laws of Kepler from this hypothesis. Though Newton had proved these properties without difficulty, he did not bother to inform Hooke of his results—a negligence that would cost him a good deal of energy during a long dispute for priority. Four years after receiving Hooke's letters, and only at Halley's insistence, Newton took the time to put down his results. In 1685, while working on his masterpiece *Philosophia Naturalis Principia Mathematica* (*The Mathematical Principles of Natural Philosophy*) he finally reached the conclusion that gravitation is a universal property and claimed that the attraction between celestial objects is mutual.[29]

Great ideas crystallize slowly, and gravitation was no exception. Newton formulated his model gradually. Two manuscripts written before 1685 show his interest and progress in both gravitation and mechanics. In *De motu sphaericorum corporum in fluidis* (*On the motion of spherical bodies*

---

[28] Westfall, p. 143, cited from the Portsmouth Collection of Manuscripts 3968.41, f.85.
[29] Gjertsen, pp. 238-239, and Arnold, pp.14-15.

in fluids) he gave a crude formulation of what would become the *n*-body problem. There he wrote:

> The orbit of any one planet depends on the combined motion of all the planets, not to mention the action of all these on each other.[30]

From this he derived the *action-reaction effect: action and reaction are equal and opposite*, which he later found important enough to take as the third law of mechanics. He stated this principle in a paper that shortly followed the previous one. There he claimed that

> As much as any body acts on another, so much does it experience in reaction.[31]

But the gravitational model failed to untangle the mystery of gravity. Newton did not say what gravity was, he only described its effect. He found the *how*, not the *why* of celestial motion. Though going much beyond Kepler by obtaining the laws of motion from a more general principle, Newton failed to explain the cause. In 1693 he would write in disappointment:

> That gravity should... act... at a distance through a vacuum without the mediation of anything else... is to me so great an absurdity that I believe no man who has in philosophical matters any competent faculty of thinking can ever fall into it. Gravity must be caused by an agent acting constantly according to certain laws, but whether this agent be material or immaterial is a question I have left to the consideration of my readers.[32]

Today, more than three centuries later, we still do not know what gravity is.

6 A *Smoother Pebble*

The plague years were a beginning in Newton's life. Recognizing the young man's extraordinary talent and contributions while becoming himself more interested in non-academic issues, Isaac Barrow resigned his chair at Trinity in 1669 and arranged that Newton be elected Lucasian Professor of

---

[30] Herivel, p. 301.
[31] Herivel, p. 312.
[32] Turnbull, Scott and Hall, Vol. 3, pp. 253-254.

Mathematics in his place. This prompted the right response from Newton, who dedicated much of the next decades to the understanding of the topics he had outlined in the Notebook. Beyond his work in mathematics, physics, and philosophy, Newton also spent much time and energy studying alchemy and religion. Many of his writings have a pious cast.

Newton's reputation grew beyond the usual proportions after the 1687 publication of *Principia*'s first edition, but in spite of the fame he enjoyed, he was lured by some of his contemporaries into many minor intellectual disputes and conflicts. In 1696, his last year at Trinity, Newton agreed to bring out a second edition of *Principia*. But it would take him seventeen years to make substantial changes and prepare the final publication.

In the meantime he resigned his chair in Cambridge to accept a job with the Royal Mint supervising the minting of new currency. In 1705, two years after he was elected President of the Royal Society, Queen Anne knighted him in Cambridge, showing the Crown's recognition for his contributions to science, philosophy, and public life. But this didn't help him succeed at the highest level in politics. The same year, he lost in the Parliamentary elections.

Newton died on 20 March 1727 and was buried in Westminster Abbey. Since then his fame has grown steadily, unsurpassed by the deeds of any other scientist before or after him. As if predicting this ascension in history, his tombstone inscription reads: "Mortals, congratulate yourselves that so great a man has lived for the honour of the human race." Though aware of his value, Newton had never dreamed of being so highly regarded. Towards the end of his life he wrote:

> I do not know what I may appear to the world, but to myself I seem to have been only like a boy playing on the seashore, and diverting myself in now and then finding a smoother pebble or a prettier shell than ordinary, whilst the great ocean of truth lay all undiscovered before me. [33]

A group of pebbles formed his theory of gravity. Among them was a pile he did not name, which, some half a century later, Laplace would call *celestial mechanics*. And one particular pebble has looked prettier than all the others, at least in the eyes of those who have tried to untangle its

---

[33] Bell 1937, p. 90.

mysteries. It was named the *n-body problem*—the main question in celestial mechanics. For the last three centuries many famous mathematicians have researched it. Thousands of papers and dozens of books have been written on this subject. In 1904 the prominent British mathematician Edmund Taylor Whittaker characterized it as "the most celebrated of all dynamical problems." Several branches of mathematics sprouted from it, one of the most resonant being *chaos theory*. With the help of celestial mechanics humans landed on the moon and started the exploration of the solar system. All those who made these things happen stood on the shoulders of Newton.

There was no magic star on Christmas night at Newton's birth above the house in Woolsthorpe, no prophets predicted his arrival into this world, and no pilgrims came from far away to greet him. But he grew immortal through what he gave to humankind. A 19th-century painting by Benjamin Haydon, today in the Philadelphia Art Gallery, shows Christ's entry into Jerusalem. Newton can be spotted in the crowd.

Bibliography

V. I. Arnold, *Huygens & Barrow, Newton & Hooke*, Birkhäuser Verlag, Basel-Boston-Berlin, 1990.

E. T. Bell, *Men of Mathematics*, Simon and Schuster, New York, 1937.

E. T. Bell, *The Development of Mathematics*, Dover, New York, 1992.

J.M. Child, *The Early Mathematical Manuscripts of Leibniz*, The Open Court Publishing Company, Chicago, London, 1920.

W. S. Churchill, *A History of the English Speaking Peoples*, Dorset Press, New York, 1990.

Collier's Encyclopedia, MacMillan and P.F. Collier, New York, 1990.

F. N. Diacu and P. Holmes, *Celestial Encounters—The Origins of Chaos and Stability*, Princeton University Press, 1996.

M. Feingold (editor), *Before Newton—The Life and Times of Isaac Barrow*, Cambridge University Press, Cambridge, 1990.

D. Gjertsen, *The Newton Handbook*, Routledge and Kegan Paul, London and New York, 1986.

J. Herivel, *The Background to Newton's Principia. A Study to Newton's Dynamical Researches in the Years 1664-84*, Claredon Press, Oxford, 1965.

A. E. Hurd and P. A. Loeb, *An Introduction to Nonstandard Analysis*, Academic Press, New York, 1985.

Newton's manuscript number 130.4 in the *Keynes Collection*, King's College Library in Cambridge, England.

A. Koyré, *Newtonian Studies*, Chapman and Hall, London, 1965.

H. Pemberton, *A View of Sir Isaac Newton's Philosophy*, London, 1728.

Petit Larousse Illustré, Librairie Larousse, Paris, 1991.

A. Robinson, *Non-standard Analysis*, Proceedings of the Royal Academy of Sciences, Amsterdam, 1960.

Y. Ryabov, *An Elementary Survey of Celestial Mechanics*, Dover, New York, 1961.

D. E. Smith, *A Source Book in Mathematics*, McGraw-Hill, New York, 1929.

J. Stillwell, *Mathematics and History*, Springer Verlag, New York, 1989.

H. W. Turnbull, J. F. Scott, and A. R. Hall, *The Correspondence of Isaac Newton*, vol.1-7, Cambridge University Press, Cambridge, 1959-1977.

R. S. Westfall, *Never at Rest—A Biography of Isaac Newton*, Cambridge Univ. Press, Cambridge, 1980.

W. P. D. Wightman, *The Growth of Scientific Ideas*, Oliver and Boyd, London, 1950.

Adam Dickinson

The Ghosts of Departed Quantities

Speak of small,
so small
that we differ from it
by as little as one wish.

But this implies motion,
sidling up to hope.
Are we any closer?

Cantor's family moved from Judaism
to Protestantism,
from Russia to Germany.

To prove that two sets of objects are the same size,
Cantor relieved us of counting.
Simply put your hands together,
no need to add the fingers.

Relation before number;
body to faith.

A set of misunderstandings,
a list of languages you never learned,
the number of intentions two people have in common,

all the days that have come before today.

He put the infinite in a bag
and shook;
everything came out larger.

What separates us is innumerable,
but like applause,
other words for the same thing,
all our differences fit.

Eclipse

Anecdote is the lowest form of evidence
that medicine will consider.

But stories grow out of the sick;
narratives hatched from the generative eggs
of endings.

There is a psychological condition
where people cannot be kept
from looking at the sun.

Night is suffocation
by the planet's
weight and width.

We are more like plants than we care to admit.
Rain is exchanged in our hair
as it is in the pine and broadleaf crowns.
We hunch down; the business of water
takes us into our bodies,
old messages stem in the spine.

A plant, a person,
is the small story of a place,
an anecdote of acid or lime.
Larger conversations of boreal and palm
contain such dialects of shadow,
rock face blocking the sun.

Solar complex. Solar plexus.
Freudian slip.
What is wrong with us
gives itself away in broad daylight,
in the very words that grow
from hands to mouths.

As in Latin,
disaster: the unfavourable aspect of a star,
a great or sudden misfortune.

Contributions to Geometry
The Gulf Stream

Hot : Cold

People in fires
crowd exits.

Lost hunters at night
climb inside the stomachs of deer.

Form is really
just accommodation.
The four dimensions
are four climates,
each with their own pressures,
and furniture.

Conveyor belt, heat pump, ribbon.

The sine wave of the seasons
has not always been so neat.
With every ice age the current slows, negotiations fail.
Tropic doesn't speak to pole.

Long before Columbus, strange woods and fruits arrived on the
     shores of Europe.
Hurricanes stayed alive until Ireland.
Postmaster Benjamin Franklin observed that the sea off Nantucket
did not sparkle at night.

Hot water and cold
are the accidents of organism.

Ice is not modern.
Our clothes become part of us.
Climate is a haircut or a shoe.

When the stream moves over the undersea mountains off New England,
the bottom is sheared off and giant gyres spin out to the east.
Beginnings resemble each other.
There is an exit at the front and the back.

Great Chain of Being

Linnaeus connected the world through teeth,
beaks, and bills.
This was the point where one thing entered another:
minerals the appetite, voice
the open air.
Ornament entered function.
And so it was that the vernacular languages of Europe were insufficient.
Only Latin or Greek;
other tongues were dark and crowded.

Chokeberry seeds must first pass through the intestines of black bears. In abundant seasons, a sow comes upon the patch on the open edge of a riverbank. While chewing, something provokes her to turn suddenly upstream and brace herself against a rock. This small amount of anxiety stimulates enough acid in the stomach to break down the hard shell of the berry seeds.

The mouth is the symbol for a corner.
Phoenicians built the alphabet out of joints,
sounds whose shapes in the throat and lips
were translated into sticks
piled or bent on the page.
Small fires grew.
People stood at windows to watch,
arms outstretched.
For the Greeks this was epsilon;
for the Romans the letter E.

Cough then glottal stop,
heartbeat then iamb,

marshland then coal
then greenhouse.
All bells are held at the top,
just as all plants are tied to the sun, just
as language, despite its vacuums and cinder blocks
hangs above heads,
rings in the ears.

Sometimes in Canadian forests small pale plants stick up in clusters from the ground with flowers that hang from their tops like bells. These are pine-saps, ghost plants, or corpse flowers. No Kingdom will accept them (neither plant nor fungus) because they are vascular organisms that do not need the sun: plants with no chlorophyll, mushrooms with rudimentary leaves. Aboriginals pick *eyebright* for the eyes; European settlers, *convulsion weed* for the nerves.

Linnaeus read the atlas wrong
and gave plants in the high Andes
names derived from the arid New Mexican plains.
Nature doesn't jump.
Kingdoms are carefully spaced ladders
against the sides of burning buildings.
One rung at a time,
women and children are the first to descend.
If the final goal of creation is us,
then why for the index of berries in a small pamphlet
did Linnaeus write: "too sour"; "black and unpleasant"?

Unlike most birds, he said, swallows do not migrate during the winter months to warmer southern latitudes. Instead they gather in the late fall at the margin of cold lakes and estuaries. Here they plunge themselves over the edge of the ice and pile on the bottom like hibernating frogs. If you come upon a lake full of swallows and break the ice in the parts that are darkest, the birds will appear in their masses, cold, asleep, and half-dead. If you fish one out and warm it with your hands, it flies away too soon. Every hole beneath it mistaken for an opening.

Homo sapiens was a draft.
So was *Homo diurnus*.
Both were crossed out and re-inserted.
Well before Darwin, Linnaeus put us in with apes;
the only difference he could see
was in the canine teeth.

Whatever is, is right.
This is not an order but a riddle,
not a single thought, but many.

Susan Elmslie

Algebra

from the Arabic al-jabr, *reunion of broken parts*

Let us for a moment say we need
to work on the rational and radical expressions of our need.

Let one side be *y*
and the other *y*,

to open the question, redintegrate
the vulgar fraction of one above the other—

where the power imbalance starts.
Let us learn this algebra, reunion of broken parts.

Chemistry

for Wes Folkerth

*and to the finer chemistries
that make up, and renew him,
every seven years, exactly
as he is, affirming everything.*
— Bronwen Wallace, "Change of Heart"

This year makes six since we crossed over, into
marriage, this ordinary brownstone, this house
made of hummingbirds. Seven, since we first
touched palms, ignited, lived close to the flame.
That we've managed to stay there, warm, sane,
neither of us succumbing to disease,
seven year itch, the lure or curse of statistics,
stills me. When he returns from this long journey,
we'll raise a glass to new discoveries
and to the finer chemistries

that kept us, keep us, coming back for more:
the sublime stitching of energy and mass
the distant tungsten burners of the night
witness and bless; the perfect resonance
of two complex structures, ours.
More than once, fully awake (the twin
engines of our cats in the duvet clouds
around us), I've studied the poetry of
the cells, elements—carbon, oxygen ...
—that make up, and renew him,
this man who was born. Who breathes beside me
every night. Basic elements, standard

measurements. Beyond this, he cannot be glossed.
What makes us who we are, all shoulder or art?
How do we learn to love what we need?
Not volatile, stable. Gentle symmetry.
Wonderingly, we're content to watch the light
show of unresolvable substances
repeating their elemental glory
every seven years, exactly.

My first lover tried to choke me. Seven
years we were together, and for seven years
after leaving him I smoked like a chimney;
took that part with me. Dark
alchemy. I transmuted what pain still lived.
Maybe love is the hardest discipline.
Simplest formula, arcane catalyst—
who can name all its allotropes?
This climacteric I'm remade, glittering
as he is, affirming everything.

Claire Ferguson

Eine Kleine Rock Musik III

photograph by Ed Bernik

This sixty-pound piece of honey onyx was quarried by the sculptor in Utah, where prospecting is not an uncommon avocation. Helaman was guided by two fellows who had modified a four-wheel-drive truck with two fifty-five-gallon oil drums. Each weekend for ten years, the two had driven into the Great Salt Lake Basin to prospect for uranium. Their method was to drive until one drum ran out, then turn toward home. Over the years, they had found everything but uranium—including an outcropping of onyx—the end of a barrel-shaped vein crushed by the weight of a mountain.

Inspired by the mathematics of a Klein bottle, Helaman sculpted the quarried stone into Eine Kleine Rock Musik III. A Klein bottle is one of four fascinating ways to sew up a square. Actually, it is not a bottle at all, but a non-orientable surface—two moebius bands joined together. On such a surface a right-handed bug would go for a walk and come back left-handed because it would actually be flipped over without realizing it. The name "bottle" was mistakenly assigned to the surface because the German word for surface, which sounds like the word for flask, was mistranslated as "bottle." Ironically, the sculpture looks vaguely like a pitcher without a spout—perhaps an avant-garde design in which a cleft replaces the nozzle.

A Klein bottle is a self-penetrating form—its surface passing into itself and re-emerging on the other side. The cream-veined salmon onyx is translucent in the line of the cross-cap or cleft—the emerging light provides the focus of the piece counter-balancing the negative space of the hole. As the artistic spirit works by contrast or counterpoint, so this sculpture marries the analytical to the subjective. The Klein bottle is conceptual; it doesn't quite exist in real physical space, but Ferguson's Eine Kleine Rock Musik III has brought it quivering to the threshold of reality.

Wild Singular Torus

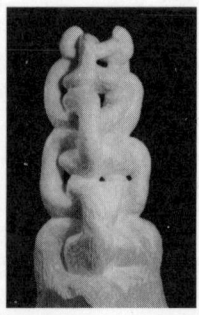

photograph by Mark Philbrick

A little over five feet tall, Wild Singular Torus is a one-thousand pound column of white Carrara marble, carved with a rope-like texture into an intricate series of knots. It stands as a solid mass of curves and hollows, like Rodin's Balzac—only a few penetrations make their way entirely through the solid stone. The idea that the sculpture projects into reality is that of two strands, infinitely intertwined upward. That being impossible in physical materials, however, they are severed prematurely at the top, exposing the raw rock. If allowed to continue knotting, the strands would continue like Zeno's paradox, approaching but never meeting a point about two feet above the broken rock.

Mythology and religion traditionally supply the symbols which bear up the human spirit, providing subjects for artists. As a symbol for continuous creation, Wild Singular Torus speaks of Earth's navel—

alluding to the mystery of the maintenance of the world through the continuous miracle of vivification, which wells within all things. It reflects a mythology of science, a concept of continuous growth, of infinity.

As a weaving, the rock recalls a prehistoric activity undertaken in the protection and fellowship of the hearth, its stone fabric symbolizing the intersection of lives against the threat of death which will ultimately rend them apart. Helaman's sculpture conveys a permanence and stability which reaches into the viewer's space by the power of its sheer mass.

Torus will endure well beyond the fragile present.

Igusa Conjecture

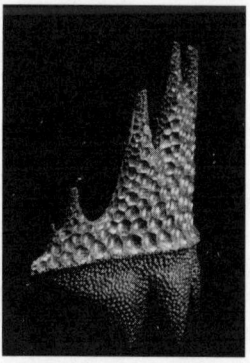

photograph by Ed Bernik

Theorems are created or discovered, but before they are proven, rationally organized, demonstrated and properly stated trophies of the mind, they live vibrantly as conjectures, teetering precariously on the ridge between true and false, proven and unproven, provable and unprovable

As navigation determines the course of a seaworthy vessel, long-standing conjectures have an important impact on the direction of mathematics. Jun-Ichi Igusa's conjecture relates two constructions long considered to be very different. This sculpture, as the conjecture, is divided between poles and eigenvalues. The occurrence of top and bottom graph points enunciates the conjecture of Igusa. A masted ship with many rudders, its sails a true course of unrealized implications.

The sculpture is approximately triangular in shape in two ways: if the stern of the ship is the apex, a vaguely triangular form can be overlaid from that point to the base. As one moves away from the prow along the esker which bisects top from bottom, volume increases with the width, providing a second triangle. The prow is thrust forward, racing eagerly toward the future. There is a sense of anticipation and excited animal energy, as though a friendly fire-breathing dragon inspired the vessel—the likeness of its snout carved into the overhanging prow, masts held aloft by its dorsal spikes and the hold of the ship morphing into its legs. Along its back the scales are larger as protective armor, catching the light in a hundred gleaming points as a thousand tiny jewels shine from its legs and unprotected underbelly.

Picasso said he did not search for new forms; he found them. Helaman finds the theoretical spirits of new forms inhabiting his mind and embodies their essence.

Emily Grosholz

Hourya

The enormous, high-ceilinged apartment near Trocadéro
Echoes, though it is full of books, intaglio'd furniture, and flowers,
As if reflecting the old house in Rabat, now seized and lost,
And the great, oceanless dunes ranged beyond the city walls
That bear the trace of wind sifting, but not of mind.

You write the history of mind, entering its formal labyrinth
With only the silk thread of demonstration to lead you on.
So Hilbert guides you, Poincaré, Weyl, Noether, Cavaillès.
So Emmy Noether grieved for Hilbert's house, her home and circle,
Stranded on the outskirts of Philadelphia, where she died.

So Göttingen fell, the greatest commonwealth of mind
Europe ever knew, dismantled by the agents of the Reich
Who sized up living mathematicians as Catholics, women, Jews.
So Cavaillès was shot against a wall, so Emmy Noether,
Exiled from her algebraic home, succumbed to memory. *Don't you.*

Trying to Describe the Reals in Cambridge

*"For there are two labyrinths of the human mind,
one concerning the composition of the continuum,
and the other concerning the nature of freedom,
and they arise from the same source: infinity."*

—G. W. Leibniz, "On Freedom"

Draw the curtains! The curtains are always closed
on roses, rugby field, light variable
but waning along these tiered northern skies
where ten o'clock's the apogee of day,
a full moon pewtering the cliffs of sunset.
I write in the wizened glow of my computer.

I write, the reals are really not like numbers
that we are used to count with, to begin
and go up stepwise. They are number flooded
by continuity, the line upbraided
by differential strands to labyrinth.
They are the shape and cardinal of freedom.

Abysses along abysses along abysses,
yet perfectly defined. As if we charted
a finest-grained Grand Canyon with passing walls
through which a sourceless unplumbed river ran,
like moonplate cumulant in tiers above
the river of waning sunlight. Draw the curtains!

Lauren Gunderson

The Ascending Life

The 12-inch Cassegrain telescope sat fat in Frank's yard pointed to the sky. Frank was ready, bending the eyepiece to see the night in its true star-flung luster. He cast the scope slowly scanning the firmament. The 8 pm sky was a heavy blue, mottled with thin clouds and high pricks of light. It was beautiful. But Frank couldn't take *beautiful* to his wife and give her any sense of being here. He couldn't do anything for Irene now. Just wait.

Frank turned to the east and a bit more to the south searching for her constellation, knowing it was dim and easy to miss. The Crab. Twenty-three degrees in the winter sky. Rising southeast. Eight stars. Soon to be nine.

Paul made noise as he opened the back door.

"Dad."

Nothing.

"What are you doing?"

"Praying."

"It's freezing outside."

"That's why I'm praying."

"Dad."

"Yes."

Frank didn't look up from the eyepiece. He knew what Paul would look like—standing, arms crossed, head cocked, lips tight. He'll snort like a kid just like he used to, Frank thought. Nothing much changes. Frank thought of the tired image of his wife, thinned into a paper version of her self lying in their bed, hooked up to dripping machines, fading slowly, surely. Everything changes.

Winner 2004 Norumbega Fiction Awards: Eaton Award for Short Story as "Cancer/Dish"

Paul blew hot air out of his mouth, a boyish gesture. Paul was a good son, the kind that followed in his father's footsteps complaining all the way. At 60 Frank realized that they were exactly alike. They both ended up astronomers, teachers, members of the same faculty for God's sake. Both taking care of a mother, a wife, a woman in terrible pain. For God's sake.

"It's cold," Paul said.

"It's important," Frank said.

"Don't be weird, Dad."

The night was quiet and clear. It smelled like old wood, dirt. Frank felt righteous, sure, eager. He knew she'd be coming, he knew she was close. He perused the sky through the telescope, passing stars, star clusters, and gas nebulae. He could name every object, every pattern. Irene, on the other hand, could name every story, every character. When Frank used an atlas, his wife used a myth. He'd always loved telling it that way. The most ancient tales infused with the latest technology; stories and science, Irene and Frank.

It was Frank's greatest achievement, steeping his wife in astronomy. It was just a brochure and a whim 20 years ago, but it was the embracing of her strength. He remembered her strength. Her insistence. God, he loved it. Making sure Frank got the telescope he really wanted, not the cheap one that would surely disappoint. Declaring never to cook that same dish twice, unless a guest joined the regular diners. Ensuring her bed was moved to the window to better see the sky. Demanding to die at home.

"Well," Paul said, boyish again, "what are we looking for?"

"Your mother," Frank said. "She should end up right about there."

Frank pointed to her constellation. It was the only way, Frank thought. The only way it could work and be beautiful. Just like her stories, like her myths, her people held on high forever. Irene would join them. She would be lit.

Paul snorted.

"Dad..." he said.

"Your mother and I have already discussed it. Help your brother inside."

Aaron, the youngest, was a chef. Inside this family home Aaron was cooking spicy things for his mother. She couldn't keep anything down, and wasn't conscious long enough to try, but requested the smell just the same. Curry and ginger, steak and fish, buttered breads, apple pies, and coffee filled the house; everything un-eaten. He loved her much more than furious baking and frying could mask; he cried while he kneaded.

"Mom's slipping," Paul said. "Within the hour. Can you stop acting crazy and make her last moments about her. Please."

Frank looked away from the eyepiece. He looked directly into the eyes of his son, so like himself it was as though a mirror was showing him his youth. He saw himself standing cold, hurt in the night and said, "I'll wait for her here."

∞

Irene's Thursday meal has vanished. The chicken shredded, peas smashed, potatoes smeared. It's okay. Happens all the time. A matron's duty. A hungry house. As Frank and the two boys finish all but the grease, Irene eats a Powerbar on her way to Emory University Adult Astronomy class, room 804, Candler Building, Milky Way Galaxy, our Universe.

This is the third class of this course and (counting a one-day lecture sponsored by Southern Living Magazine, *Easy Thanksgivings from Turkey to Trimmings*) the fourth collegiate-level class of Irene's life. She never went to school; went straight to marriage, kids, soaps at 3, dinner at 6. Irene doesn't regret anything and lives happily in her earthly patterns. She is grounded. She is mild.

Tonight Irene and her classmates get to know Jupiter. Having introduced themselves to Mars and Venus last week, and having mingled with Mercury and the Sun the week before, room 804 is getting crowded with friends of the solar system. Professor Callan is a great host and supervises the celebrity encounters with the style and energy of a Parisian Salon. Everyone entertained. Everyone intrigued. Everyone leaving with celestial gossip. Mars has ice caps. Venus is way too stuffy.

As Irene takes out a pencil from her heavy purse Callan introduces the first slide of the night. Jupiter glows. Like marble-orange caramel, warm and brave the planet roosts titanic on the wall. Irene smiles. Jupiter reflects in her glasses. The company of this giant body swallows all but the light-speed desire to know its every curve. Callan shows the class Jupiter's red storm (... constant), moons (... violent), and ring (... thin as ice on edge).

Irene is in love, blushing, splashing air on her happy cheeks. The planet reminds her of her husband, at home right now cleaning dishes from this Thursday's chicken dinner. Frank: a good, solid planet-man.

In a moment of fast flirtation Irene thinks she'd like to try mashing sweet potatoes and white potatoes and swirling them together with candied walnuts. Yes. And maybe caramel apple crumble with melted ice cream. Cranberries. Things with salsa, cayenne (... red storm).

As always class is over too soon. The whir stops. The titan leaves.

"Inspiring, Professor," Irene says, "Where can I get more?"

"Info?" he says.

"Jupiter," she says.

Callan gives her some magazine titles she's sure her boys can find. Her boys: her comets, fast but predictable, no damage, lots of dust.

Next week is Saturn preceded by the same Thursday chicken, peas, and potatoes. No. Not the same.

"What's in the chicken?" Frank says kindly, casting no judgement, just noticing.

"Cayenne," Irene says, giving nothing else away.

Her boys don't care for cayenne, and eat too many cookies as Irene drives off to Saturn. Frank puts the chicken in air-tight plastic for lunch tomorrow.

Saturn (... hearty, mild) offers Irene custard thoughts. Custard with white and dark chocolate rings. Cilantro polenta with créma and green-chilé strips. Baked pear in spiraled sweet cream. Uranus and Neptune (...cool, huge) inspire cucumber dill chutney with pan-seared pepper steak and ginger jelly. Pluto (... slight, crisp) suggests plum sorbet on frozen coconut thins. And on and on through the universe. Salty comets, bitter quasars, sweet and sour nebulae. Irene's notebook, borrowed from her boys, overflows with potential, impossible, celestial dinners. She's never thought like this, never dreamed outside herself. God bless the Galaxy.

Resting her brain with Hamburger Helper three nights in a row, Irene finds mid-week inspiration by checking her husband's *Sky and Telescope Magazine*. Her boys don't mind boxed dinners, Frank thinks it odd but not inconvenient, and Irene is on a different planet. Her men are fed, her mind free, her stars hot. It's Thursday again and she can think of nothing but the universe. She leaves for campus, racing Einstein's beam of light. Nothing faster.

Finally, as the class rounds the Milky Way and heads home (... green planet three) Irene speaks.

"Would y'all like to come over to my house for dinner and some star watching?"

The class is thrilled and free next Thursday.

As soon as her boys leave for school, Irene is shopping. In and out, up and down the aisles of the Farmer's Market she moves frictionless in her trajectory. Spices she's never heard of, fruits seen in postcards, vegetables as bold and heavy as gold. Breads baked in shapes not square. Fish and cheese bought outside the plastic. Everything thrown into her certain orbit. Those savory solar winds.

The cooking commences Thursday morning. The planets align. The signs are clear. Chop, coat, fry, bake, cool, stir, rise, chill, roast, orbit, orbit, orbit. Her star is being born (... bright, embroiled, heavenly.)

Callan comes early to set up the telescopes. Irene's oldest, Paul, who would eventually teach Astronomy and Astrophysics at Emory, helps Callan carry the precious equipment. Paul is delicate and proud to handle these machines of vision; he'll remember this.

"I brought red and white wine," Callan says kindly, already impressed (... *those savory solar winds*). "Didn't know if we'd be having chicken or beef."

"We're having both," Irene says leaving a comet-tail of flour and garlic as she returns to the kitchen. "It's perfect, Professor. Thank you." Irene's voice trails off into her mixing. Callan heads out to the patio.

Frank appears in her kitchen. Irene turns and sweetly touches his cheek, she loves him and she was handling caramel. She can tell he is proud.

"You're a star," Frank says.

Irene knows it's a swelling compliment.

Everyone arrives on time and enjoys Callan's mid-appetizer lecture on tonight's constellations. Irene's youngest, Aaron, who would be head chef at the Peachtree Ritz Carlton, passes around her appetizers—stuffed shrimp, Cuban bruschetta, roasted mushrooms—Mercury, Mars, Saturn. He enjoys the job, the grateful grabs, the easy pleasing; he'll remember this.

Jupiter is served for dinner—beef medallions in curry pools topped with fried yam curls, marinated jicama and pomegranates. The company applauds the banquet. The entrée makes a decided dent in space-time. The class of Emory University Adult Astronomy chats, indulges, and explores the world in front of and above them. All patio-passions collide tonight—philosophy, gastronomy, and looking up in awe.

But dessert eclipses. White-chocolate Neptune with raspberry crème and candied constellations.

The back yard is happy, the sky is full, the smell of Irene's home is finally right (... *those savory solar winds*). The stars and Irene's fingers smell of mint and brandy; both coruscate in the night – the stars because that's what they do, her fingers because she's licked them sugarless. Irene smiles. The night-immensity reflects in her glasses. The universe cools and expands.

∞

Frank was cold, his fingers revolted into thick sticks that made subtle maneuvering of the Cassegrain clumsy. He stopped to rub his hands together. He wished he was dying instead of her, or *with* her, either way would be fine. Not this way. Not this sly slip, this erosion. Beyond unfair, he thought. Unjustifiably cruel.

Frank had held her every night since the relapse. He rubbed her feet that didn't touch the floor anymore. He played her music, sometimes just whispering *You are my sunshine*. He wrote down recipes as she imagined them for Aaron to try sometime. He'd moved her bed to the window so she could see the sky. She'd insisted. Frank relied on this insistence. It was how he knew she was still there. Frank spent the last few nights out here because Irene was fireless, and Frank was bare without instruction. She had told him watch the sky, to find her a seat.

∞

Irene's oncologist is speaking to her in a cool white room on the fifth floor of a very good hospital. Irene is diagnosed—*it's deep in the bone but we've got some options*. Irene is oriented—*start chemotherapy immediately*. Irene's cookie crumbles—*too much flour, over-baked, smashed under God's heel.*

I don't care for God, Irene thinks wishing she was in one of her myths. Zeus, Athena, Apollo; at least you can count on emotion with the Greeks. She could plead, or bargain, or sacrifice something. But God, this singular, heavy nowhere-man letting this all play out without regard or rhetoric. Who is this guy? (... *sour air*).

Frank tightens the hold on her hand as her world tilts on its already wobbly axis. The precession of human life, the joy-sorrow circle continuing for eons, has come around to her. Irene squeezes her husband's hand and agrees to go ahead with the full-chemo treatment program. Nausea, hair loss, fatigue. Whatever it takes, she thinks, kill those little bastards.

Irene's head simmers in the possibilities. Dying, living, pain, explaining to others, taking charity, asking for things. Her world has just been stolen from her, eclipsed by imminent, over-seasoned, salty sorrow. She knows she'll not win. She senses it. Crazy, she thinks, to know how it will end.

She lays her head against the seat of their car as they bumble home in the 4 pm sun. On the way Irene decides to have a party. Tomorrow. Yes. A big dinner party for all her friends; with mixed drinks, and caviar, and cakes. She'll wait till dessert to tell everyone, to say the word. She wouldn't want to ruin the crab legs. Yes. She'll serve crab: boiled to death, and made tasty with garlic and butter. She'll buy the crab whole. And cook the hell out of it.

Working through the lethargy that accompanies bad news, Irene outdoes herself this time. Winter pomegranate salad, almond crusted Brie, rose-water cream chicken. Aaron helps of course, making his specialty: wild mushroom wood sauce and biscuits. But the crest of the meal, the feature, is the plain, red-white crab legs blooming mid-table. The house smells densely of garlic, butter, and seafood. It overpowers.

As their guests arrive, Frank and Irene greet them one by one. The neighbors from both sides, some colleagues of Frank's, and most of Irene's original astronomy class find a seat in the dining room. The meal begins. The chatter wanders through children, taxes, high school basketball, and how fabulous the meal is. Aaron is sent to bring out the dessert. The guests wait tightly with the memory of Irene's other astonishing treats.

"Friends," Irene says smiling gracefully, "I invited you all here tonight, to celebrate good things. To show you good living, good friends, and good food. I also have cancer. There are some treatments, but... no one is sure quite what to expect. So. Enjoy dessert. Know that I'm fine. And keep in touch." Irene hasn't stopped smiling, and sits down to Aaron's perfect chocolate cake.

∞

Frank heard his youngest son singing. Aaron's wobbly timbre, scathed from sleepless hours and tears, slipped through brick and plaster to their yard outside.

You are my sunshine my only sunshine...

Frank understood that it was Aaron's way of summing up his mother, the woman and the memory of the woman. All the minor sicknesses she'd nursed, the meals she'd delivered, the cards, the parties, the trips, the family network she'd begun.

Frank thought a sudden thought and darted in towards the music.

You make me happy when skies are grey...

∞

Frank is very handsome, Irene thinks watching him work in the shadows of his desk. He is so determined, so raw. She sees Frank rip out a page in the typewriter and throw it to the floor.

My goodness, she thinks holding her stomach. She knows this is important. This paper could be his definition, his ticket to a professorship. She shouldn't disturb him, but feels herself gliding to his side anyway.

"Can I help?" she says.

"Coffee," he says.

"How about a sandwich? A cookie? Some brain food," Irene giggles supportively and tickles his neck.

"Coffee, please," he says not looking at her.

"Honey, it's been hours since you—"

"JUST. Coffee, Irene. Make it, buy it, grow it for Christ's sake. That's it. Dammit." His hand goes straight for his forehead as he bends toward his notes on the desk.

She waits.

That hurt very much. She hates him for a moment, then thinks that's too much to commit to. She decides she hates coffee, that's easier. Snobby, rude, coffee.

"Black coffee. Please." Frank says.

Irene backs out of the room, stumbles over a book on the floor, kicks it, and leaves. First make some bitter coffee, she thinks, then tell Frank she's pregnant, then choose a book of his to peruse, to contribute.

∞

Keep singing, Irene thinks as she dies. She breathes a last quick snuff of cinnamon air, and leaves them and the song mid-verse. Goodbye guys, she thinks; and then, in the same green moment, Hello.

∞

Irene realizes quite suddenly that her husband's field was full of stars. Not the big red or white ones, but characters straight out of the movies. Heroes, villains, love-stuck ladies, jealous husbands. Beautiful rich stories all drawn out in the sky, charming constellations existing since ancient times. Looking up at night was like reading a picture book. Irene loves that idea. She also loves that she has a constellation, a sign to which she was born, a slice of the sky. Irene decides to pay more attention to her constellation, an eight-star block supposed to be a crab.

In all this, Irene seems to complement her husband. Where Frank sees astrophysics, Irene sees a hunter and a bull. Where Frank sees explosions millions of years old, Irene sees long-tailed bears, happy twins, and imprudent queens. The language of the sky was not a mathematical ramble, but a story. And one that she wanted to understand.

"Can I see Taurus?" she asks Frank one late night while her child dozed unborn inside her.

"Sure," he says.

They drive to the telescope at the university. Frank has been given the key after being appointed an assistant professor in the wake of his article. He teaches Astronomy 101 now, and is in charge of forcing celestial mechanics on English majors.

"I heard that Aldebaran is red? It's the eye of the bull charging at Orion the hunter. How exciting it all is."

"Yes," Frank says. "It's a red giant."

"An angry bull," She says. "Makes sense he would have a red eye."

They ride through the night, Irene full of baby and other happy things.

They are young, thrilled, terrified to be what they are: married, expecting. They climb into the observatory, where no one is, and see that yes, Aldebaran is red.

∞

Paul met his father at the door.

"Dad," he said quietly.

"Yes," Frank said running past him. That wasn't a question. Somehow he knew. He knew inside and out—heart and stars. He ran past the copious number of uneaten dishes, past the off-turned stoves. He ran into the room where Irene lay wading in the watery moon lake from her window. She was still. Frank's world dissolved in the rip-tide fear of forever. He took her hand. He touched her cheek. He promised, in the quietest way, to keep her story with him always.

He turned to his sons—the one a mirror-image of himself, the other like his wife, quiet and sure. Frank felt himself wash away, tidal in the wake of all this. Frank felt Paul's hand on his shoulder, then Aaron's arms on his back. They bent together and held fast against the solar wind.

∞

Aaron has the flu. Paul has caught it already. Now Frank feels bad, too. Irene is fixing soups in the kitchen for her men; tomato for Frank, chicken noodle with twirly pasta for her boys. She dispenses liquids and medication in waves, four hours between doses. During the down time she reads *Sky and Telescope Magazine*. She hopes to persuade Frank, in his weakened state, to buy a new Cassegrain for the family. Twelve inches. Powerful enough to see what's really in Orion's belt.

"Mom..." Paul moans from the TV room.

"Mom..." Aaron echoes for momentum.

Irene ladles soup into bowls, adds a garnish she knows they'll discard, and delivers it to her boys. The TV room is a tangle of blankets and sheets. Aaron and Paul sprawl on the pull-out sofa-bed, floating in pillows and stuffed animals. Frank burrows into his Lay-Z-Boy, coughs snuffed by an old quilt.

"Everyone feeling better?" she says smiling.

Aaron and Paul shake their heads.

"No," Frank says with just enough of a whine to make Irene giggle.

She hands out soup, and kisses each of her men on the forehead. As she re-fluffs their pillows and re-tucks their blankets she sings, because it reminds her of her own mother: *You are my sunshine, my only sunshine...*

"We're not babies, Mom," Paul says as Aaron backs him up with a weak "Yeah."

You make me happy, when skies are grey...

"Honey," Frank says, "Could you call the school and tell them I'll not be coming in today?"

You never know dear, how much I love you...

"Can you change the channel?"

"His has more chicken."

"My throat hurts bad."

"Mom."

"Mom."
"Honey."
Please don't take my sunshine away.
Irene hums to herself all day, toasting sandwiches, measuring syrups, and feeling cake-rich in love.

∞

The sun was rising in a cloudless sky. The stars were fading quickly. Frank ran back outside to his telescope, mute in the pre-dawn dew. Frank was nervous. She should arrive any minute now. But with the morning so close he might have to wait till tomorrow night to see anything. What if he couldn't find her? What if she changed her mind—how would he know? What type of star would she be, how close to Earth? He didn't know, he couldn't know.

The sky eased into paler and paler shades as Frank prepared to meet his wife.

Aaron had put the dishes from last night on decorous plates in the dining room for friends who would be coming by. Paul had called the hospital to send an ambulance. Everything was in order, Frank thought. Except everything that mattered.

∞

With two boys, a busy husband, and a big house Irene has lost sight of the sky. Woven in a tight schedule of soccer, school, pick-up, drop-off, dinner time, bedtime, bath and breakfast, Irene feels proud to finish a day with time to imagine a different life. She *is* happy, just. . . spice-less.

"Honey," Frank says over their Wednesday stroganoff, "How was your day?"

"A day, you know," she says "More carrots?"

"I picked up a flyer at the office today. I thought you might like it. It's a new class, they're offering. Every Thursday."

"I don't have time for a class, Frank."

"You could. I can watch the boys."

"Frank."

"Just think about it."

"Mom's gonna go to school?" Paul says.

"No, baby," she says.

"Yes she is. If we're good, and promise to clean up after dinner, Mom can go have fun."

"Fun at school?" Paul says.

"Frank, I need to think about this..." Irene says.

"Whaddya say, boys? Give Mom a break? Take charge of our own castle?" Frank says.

"Sure," Paul says.

"Whatever," Aaron says.

"Frank..." Irene says.

"Good," Frank says. "I already signed you up. Classes start tomorrow night."

Irene looks at her stroganoff already starting to crust on the edges. She should start cleaning this up. She looks at her boys, eating, eating, eating. She looks at her husband who looks at her. Thank you, dear.

∞

At 6 am that morning Frank stopped looking up, and looked out over the eastern limits. He saw the sun. It glowed with enormous precocity —red, yellow, white. The flame buoyed in mid-air, seeming to flex and yawn. It sat low on the horizon staring Frank in the face. It vulcanized the Cassegrain telescope, made it shine like a missile in the light. It threw pink on Frank's house, his car, his front yard, and his boys as they made their way to his side.

"Hello," Frank said to his boys. He meant "here we go" or "don't leave anytime soon" but had only one word left. Frank squinted in the sun, recognizing its warmth, its honesty, its endurance.

"Hello," Frank said to his wife, resplendent in the low, blue sky that morning.

Philip Holmes

The Lines Remake the Places
Four pieces from a work in progress

> "...the world is, as it were, draining itself, in that the history of countless places and objects which themselves have no power of memory is never heard, never described or passed on."
> —W. G. Sebald, *Austerlitz*

> "It is no surprise that consciousness models are rare."
> —Gerald Edelman,
> *The Remembered Present:
> A Biological Theory of Consciousness*

ii
Intermediate values (1970-2000)

Coming from then continuously
to where I find myself, I must
have passed through every point between—
uncountably many presents—each one
gone in the instant it passed.

But what's grasped at all is grasped
because of that endless flux underfoot,
not on account of here and now. We see
only what's presently remembered,
not what's present, which will, in time,

be held only in memory. Lacking the mind's
landscape, each hour would be a blur
of weather; this one will come to mark us too,

dividing then from now, things from nothing.
Models lie, but they're all we have:

the past constructs the present, if imperfectly.
All that's lost, as now will shortly be, sustains us,
even when it holds no ground, is gone
sooner than the word's said, dust blown away,
a glass drained, its stain dried from the table.

iii
Commuting (2002)

In the tunnel the train suddenly stops
throwing us forward against each other.
Abruptly still, what was I thinking of?

No one falls. Some murmurings, excuses
and surprise: we readjust quickly when,
in the tunnel, the train suddenly stops,

waking us to reality, or from it.
A dozen conversations are unmasked
abruptly. Stilled, what was I thinking of?

To bring it back takes time. Two bulbs light
the cables too faintly for eyes to follow
in the tunnel. The train suddenly stops

our thoughts and snaps us back to the one
place, the pulse at first speeding and then
abruptly still. What was I thinking

when we entered this narrow place
dividing now from then? It all stops
again. In the tunnel the train's suddenly,
abruptly, still. What was I thinking of?

iv
Tropopause (2004)

Up here, past where the weather ends,
We can see the snakes of roads heading
down from pleated mountains, but no one
travels them, and the rivers are frozen.

A true pause: the world's dendrites quiet,
long vistas of cloud, ice crystals, snow;
the air's ever-present roar no longer heard,
flashes of sea as blue as an open eye.

Between us also the turbulent present
boils away, becoming habit, showing itself
as moderate and in the end quite bearable:
hardly an echo after all that clamor.

But if it seems that we might have risen
above everything, it's a false calm:
life spits and twists to get loose again
trapping a finger or foot in the door,

and I'm dumb with the need to speak, blood
rushing like traffic above all else, words
lurching in narrowing lanes and bends,
blocking the intersections till all stops dead.

But even at this still point, winter solstice,
when the darkening year's half waits to turn,
water still bustles under the ice, carrying
bubbles and cells, invisible parasites down,

down to the sea, far below where we sit,
not in comfort, but seated. The struggle

is never done. No more in abeyance, the world
comes back to bang and twitter at the door.

v
Once removed (2006)

Water through limestone, ripples on slate or chalk:
so much gone on the stream. Words thrown down
like a glove, voices raised out of hand;
yet the one voice still brings me back,

and carries us onward, making our place.
Words build it and hold it, forward and back
(at times, to go on they must be held back).
Wind falls at dusk where once it would rise;

afternoons shimmer among wisps of grass
beyond the shuttered eyes. Pay attention.
Tongues flicker lizard light a half life gone.
Words pack the air above my glass.

Words can take away the now, wind up
the time it takes to know what's changed. Yet,
pricked, we feel before we're conscious of it,
and speak without reflection every day.

To think at all we must rule out the most
of what we've come to see or hear.
Words clutter up the space above my chair.
The cost of love is ignorance, is trust.

Words pack the air beyond the window ledge
where dusklight rises to the tops of leafless trees.
To think at all we must accept the cost it takes
to make attention pay; to speak, or not.

The borders of the map cannot be sealed.
We occupy an ordinary spot.

Alex Kasman

On the Quantum Theoretic Implications of Newton's Alchemy

Only the skin around Rick's *left* eye wrinkled when he gave me the good news. That's a bad sign. Normally, the crow's feet around both eyes become more pronounced when he smiles. They don't make him look old, just extremely happy. But occasionally the wrinkles are only visible on one eye or the other.

I must have witnessed this twenty times during the four years that Rick served as my thesis advisor. I had also observed a few examples of these one-sided smiles during the previous year, when he taught my *Topics in Mathematical Analysis* class. Based on this empirical evidence, I developed a theory to explain this "broken symmetry." I was pretty sure that his non-symmetric, smile-induced wrinkles are indicative of a *forced* smile, one that accompanies a lie.

For instance, if one of the two bright students in his statistics class made a particularly insightful remark during class, I think Rick would have liked to have said "That's fantastic! You're a talented young mathematician. Have you considered majoring in math?" But, worried that this would bloat their egos or offend the other students, he would instead say instead "Yes, that's correct." This sort of remark was generally accompanied by wrinkles around his right eye.

On the other hand, sometimes he had to smile when he didn't feel like it. Comments like "The department has voted to let me serve as chair again starting next semester" or "Despite your low grades so far, if you study hard before the final you can still earn a good grade in this course" are the sorts of things he is likely to be saying while smiling with wrinkles around his left eye only.

So, I was a bit concerned when he told me of the job offer from Dr. Stein. On the surface, it was good news. Despite the difficult job market, especially for people in my field of non-standard analysis, I had at least one firm job offer.

But only the skin around Rick's left eye wrinkled.

"What do you think?" he asked me as we walked through the flurry of cherry blossom petals on the central quad. "Does that fit in with your plans?"

As if I *had* plans! My plan was to get my degree and get a job. If this was a job, then it fit with my plans. Anyway, I received no other job offers, and so had no choices to consider. Not only that, but it was a research postdoc at a mathematics institute! And, from what I had heard, Ann Arbor isn't such a bad place to be.

As I said, it seemed to be good news, but the wrinkles got me worrying.

It *was* a little bit strange that the job offer came without my even having applied. And, I have to admit I'd never even *heard* of the "Institute for Mathematical Analysis and Quantum Chemistry." This research center had no Web presence at all.

Perhaps they do government work that is so secret that security precautions prevent them from having a Webpage.

Perhaps the one-sided smile meant that he was saddened by the thought that I would be leaving soon.

Perhaps my theory about the symmetry breaking was just plain wrong.

Or perhaps things were about to take a turn for the worse.

Mulling over these possibilities as the train pulled into Ann Arbor, I started pulling my old green duffel from the overhead rack before we had come to a complete stop. So, when the train finally did stop, the cumbersome bag made me lose my balance. I fell on a woman talking on her cell phone. Her eyes narrowed, as if I'd tried to steal her purse, but she continued talking as if nothing had happened.

I was still apologizing as we entered the terminal building, but I was pulled away from her by a small man tugging on my shirt sleeve who meekly asked "Are you Igor, then?" Raising my hand one more time to try get the attention of the woman as she headed for the bathroom still chatting away, I recognized the futility of my desire to be forgiven and turned toward the man. My duffel slipped off of my shoulder and fell to the ground, sounding more like a dead body than the cheap clothes and toiletries it actually held.

"Yes", I said holding out my free hand, "and you are?"

"I'm Doctor Stein, from the Institute."

Well! The director himself had come to the station to pick me up. That seemed promising. My first impression of the man was a good one as well. His brown plaid shirt clashed with the blue slacks he wore, but he looked friendly. I am not sure exactly what gave me the immediate sense that the man was a genius, but I was already convinced that Stein was brilliant.

"You can call me Frank," he continued, shaking my hand. "Can I help you with the bag?"

I suspected that the bag weighed twice as much as he did, so I carried it to his green station wagon myself, and we were off.

At first, we merely gossiped about Rick. The two of them had apparently met in grad school and only barely stayed in touch since. For instance, he didn't know that Rick had married the well known number theorist Lisa Rojahn, or that he had been in a water skiing accident and confined to a wheel chair for almost a year.

We also talked about my name, Igor Stravinsky. Everyone asks about it eventually. I began the usual spiel about the famous composer being a distant relative of mine and how my parents, being very fond of music, named me in his honor.

But, as we entered the city of Ypsilanti, Michigan, I began to ask more pertinent questions about my situation.

"So, are you taking me to my apartment first? Will I be living in Ypsilanti and commuting to...?"

"No, no," he said cheerfully. "You'll be living at the Institute with me, and it's *in* Ypsi." Then, seeing that I seemed concerned about this bit of geography, he added: "Don't let that worry you, my boy. Did you know that MSRI isn't in Berkeley but in Oakland?"

Somewhat comforted, I pressed on with my questions. "I did try to read some quantum chemistry papers, you know. There was this cool one where they were trying to predict the color of gold just from the mathematics of it. The funny thing was that because the nucleus is so heavy, they got it all wrong unless they included some relativity in there. Well, I understood that much. But, the details lost me. I don't get Schrödinger equations and all that stuff about particles, yet. How much of that are you expecting me to learn?"

"Particles? You don't need to be concerning yourself with that. I can do all of the chemistry. You will be my 'hired gun', handling the Riemann-Hilbert problems that pop up."

"I didn't see any Riemann-Hilbert problems in what I read. It's too bad, too, because then maybe I would have understood it. So, could you explain to me how..."

"As I said, Igor, don't worry about it!"

Finally, I asked a simple question. "Do all of the Institute's employees live there, or just me and you?"

"Yes," he said, and crow's feet appeared by his right eye. There may also have been some by his left eye, but I couldn't see them.

"I'm sorry, I guess you didn't hear me. I asked whether you and I alone will live at the Institute or whether all..."

"I heard you. I heard you. Now, show me you're as smart as Rick says and tell me why I said simply 'yes.'"

It didn't take me long to figure it out, but shock kept me quiet for too long and Frank began to hum the annoying music from the final round of Jeopardy. He seemed to be enjoying himself.

"Okay," I said with too much anger. "I get it! It's not much of an *institute* is it, if we're the only two people there."

"It's significantly better than it was last week. One hundred percent increase in personnel! Not bad, not bad. And getting better all of the time. Aha! And here we are."

I had not been paying close attention to where we were going during the last bit of conversation, and so I was surprised now to see that we were in a residential neighborhood that had clearly seen better days. The houses were large, with fancy woodwork and reasonably good wooden siding. But, they were all in serious need of paint, and grass growing in the many cracks made the sidewalk almost invisible.

We had stopped in front of one of the better looking ones, the weeds having been trimmed back enough that you could see the steps. It had been painted some time in the last ten years. Near the porch steps was a little sign that read "Mathematical Analysis/Quantum Chemistry" in gold letters on a black background.

That pretty much confirmed my theory about Rick's smiles.

∞

My area of mathematical expertise is solving Riemann-Hilbert problems. I like to think of myself as a wild animal trainer, but rather than making lions jump through hoops, complex-valued functions on a Riemann sphere do the jumps for me. Calculus students never see these animals; the functions in their zoo are well behaved, continuous and differentiable everywhere. But, in many applications, you need functions that jump in rather specific ways at prescribed locations.

The thing is, it can be pretty difficult to make the functions dance the way you want them to, especially when the moves get fancy. Using the stuff that most mathematicians know, it would be a nearly impossible task. But, I'm lucky that Rick knows the latest mathematical gimmick, a toolbox called "non-standard analysis" that gives me numbers infinitely smaller and infinitely larger than the usual ones. We use these as treats to coax the functions to do what we want. It's still a lot of work, but it is straightforward enough, and not nearly as dangerous as lion taming!

My job at the Institute consists of using these techniques to solve the Riemann-Hilbert problems that Stein tosses at me. He carefully describes the jump he wants, where he wants it, and the *boundary conditions* that cage up the wild function away from the jump so it is always under control.

After I had been doing this for a few days, Rick called me to see how I was doing.

"This job isn't so bad after all," I told him. "The problem I worked on yesterday was really cool. You should have seen the monodromy that I ended up with! At first I couldn't think of what I could do until I used a Möbius transformation and it all fell into place. You know?"

"Yes, I know." Rick sighed and paused. "Has he talked to you about, uh, chemistry?"

"No. He told me not to worry about that side of it."

"That's good, I suppose. Well, keep in touch."

As I hung up the phone, I had to remind myself of where I was in my latest computation. It was essentially done. So, I finished up what I was doing by writing a brief note to Stein:

Remember, in the last formula you've got to think of the jump matrix as being an operator on the nonstandard Hilbert space H. Then, using your curve C we can define chi as:

$$\chi(\lambda) = I - \int_C F(\lambda)G(\mu)/(\lambda-\mu)d\mu.$$

∞

I'm not sure why Rick worried about me. I was beginning to have a great time. Living with Dr. Stein in the old house may have been rather unusual, but since he didn't care what I did with my time off it wasn't like living at home with mom and dad.

In any case, Stein approved of my hanging out on the University of Michigan campus with the math grad students. I attended their colloquia and seminars. I joined them in the U Club for beers. And I enjoyed a picnic lunch on the Diag with one student in particular, with high hopes of seeing more of her.

Dr. Stein (I still cannot quite bring myself to call him "Frank") wanted me to stay connected to the mathematical community. He also wanted me to report back anything that anyone said about him. And I did. I told him that his early work in mathematical physics was apparently still respected, but that he is now considered quite the crackpot. I told him that the department was trying to figure out how to fire him despite his tenure. And I told him that people were very curious to know what we were doing at the Institute.

Since he had not sworn me to secrecy, I was comfortable passing information back the other way as well. I told my friends in Ann Arbor that Dr. Stein was definitely not the craziest person I'd ever met. As far as work goes, I told them he would frequently give me specific Riemann-Hilbert problems to solve, and he was always very grateful and positive when I found solutions quickly. I had no idea what he was doing with my solutions or what it had to do with chemistry.

On our first official date, Becka and I caught a showing of *How to Steal a Million* at the Michigan Theater. We had seats right up front by the pipe organ, and in the scene where Audrey Hepburn and Peter O'Toole are locked together in the closet at the art museum, her knee rubbed gently against mine and it stayed there.

Everything was going great. But later, over coffee, she started asking me about my job again. She asked the same questions she had asked at our picnic lunch the month before. I gave her the same answers. Why I should care about Franklin Stein and his micro-institute when all I could think about was her? She didn't agree.

∞

"It looks nice," Dr. Stein was saying. "Yes, very nice. I think we're getting quite close to our goal now, boy. It won't be long now until they see that I'm not such a 'crackpot' after all! Just one question, are the branch cuts double ramified along the..."

I was barely listening. Instead, I was looking over his shoulder at the paper he had been working on. Below some complex analysis that I recognized, large and bold, I saw the expression:

$$\oiint \odot + O(\male) \leftrightsquigarrow \hbar + O(\female) \ .$$

I was thinking, "What is this, *astrology?*"

"What *is* our goal, Frank?" I found myself shouting. I had never taken him up on his offer to call him by his first name before, so he knew right away that I was annoyed.

"I will tell you, son." He spoke calmly, in contrast to my agitated state. "You seem pretty bright and may be able to understand the importance of this work more than most of them. But first, are you sure you want to know?"

Ignoring his polite response, I continued yelling things I hadn't realized I had been thinking. "You know, I took classes in quantum physics and basic chemistry in college. I read a few papers on quantum chemistry before coming here. And, I never saw *any* Riemann-Hilbert problems in any of it. Are you even doing anything with the answers I'm giving you, or do you just go off and come up with another problem for me to work on? Is that what all of this is about, just wasting my time?"

"We are not wasting anyone's time. We are doing extremely important work here, Igor. The course of history will change." His voice became deeper and very nearly *boomed* as he said "After these many hundreds of years, Isaac Newton's dream will become a reality. By piecing together a bit of this field with a smidgen of that theory, we are resurrecting the greatest scientific achievement of all time, which died a slow and painful death from neglect and misunderstanding."

He would have gone on with this declamation if I hadn't butted in sarcastically with "What? Did Isaac Newton do quantum chemistry?"

"In fact, he did. Yes, Isaac Newton invented quantum chemistry as far as I'm concerned. Of course, he would not have called it that."

I waited for him to continue, but apparently it was my turn to say something. So, after a reasonable silence, I went ahead and set him up as he wanted. With obvious skepticism, I asked "Alright, you win: What would Newton have called it?"

"*Alchemy!*"

∞

It was rude to walk out on Dr. Stein without a word, but I needed some time to think.

Alchemy, huh? That's what we're doing at the institute: turning lead into gold! I couldn't tell my friends at the University about that, and I suspected that it would not look good on a CV. Could I get hired, I wondered, at a real math department after spending a year doing witchcraft?

Mad and worried, I turned around and walked out of there. But, I had grown to trust Stein a little bit, and to like him as well. So, I wanted to at least consider the possibility that he knew what he was doing.

That's why I didn't immediately get back on a train to home but instead went to the Grad Library and did some reading about alchemy.

I was surprised to learn that Isaac Newton actually *had* studied alchemy! (One point for Stein.) I had too much respect for Newton's mathematical discoveries to discount him as another complete wacko. On the other hand, everything else reinforced my impression of alchemy as an early pseudo-science based more on wishful thinking than scientific rigor. The experts agreed that people who believed in alchemy back in the 17th century were basically crackpots, and anyone who still believes it today is plain nuts. (By my reckoning that made it two to one against poor Dr. Stein.)

I intended to pack my bags and head for home, but I saw the light on in Dr. Stein's room and knocked lightly on the door.

He seemed to know that I was planning to leave.

"Don't go now, Igor. I'm... We're so close!" he pleaded.

"You can do it without me, Dr. Stein. You wouldn't want me around anyway if I don't believe, would you?" I thought that perhaps, like psychic powers and homeopathy, alchemy could be something that never seems to work when a skeptic is around.

"I can do almost all of it without you, and have been for years. But I never could understand that nonstandard analysis that you do."

People are always saying that to me; I don't understand why. "It's easy," I assured him. "Just pretend that there are real numbers that are infinitely big or infinitely small and do what you always do in a calculation."

"It is easy for you, and that is a gift. I would be ever so grateful, if you would share that gift with me just one more time. I think I've got it now. Solve this problem for me and I will show you that I do know what I'm doing."

Pity is not one of my favorite emotions, and I resented him for playing on it. But, his ploy worked. I stayed awake that night deriving complex functions with prescribed jumps for my boss one last time.

I probably was not looking my best when I wandered into the kitchen at 6:30 in the morning where I found Stein waiting for me with a fresh pot of coffee. As I struggled to stay awake and dunked my donut, he tried to explain his idea.

"What did Newton do immediately before he turned to alchemy?" he asked rhetorically. "He defended Gassendi's particulate description of light by showing that gravitational effects create a sort of wave-particle duality. But, then Newton had an idea that nobody else would think of for hundreds of years: what if all matter and not just light could be wave-like in nature. In modern terminology we would say that he had a soliton model of fermions induced by the nonlinear effects of gravity coupled to the Hamiltonian so that. . ."

"You're saying he found a theory of quantum gravity?" I said between sips. "A 'Theory of Everything', like physicists are trying to discover today?"

"Yes, but he had quite an advantage because he didn't spend hundreds of years believing in nonsense first."

"What nonsense would that be?" I inquired, thinking for the first time, that Dr. Stein was as unhinged as my friends said he was.

"Particles, my boy! The idea that matter is made up of particles is a terrible mistake our physicists made early on and we've been paying a price for it ever since."

"Of course matter is made of particles," I said patronizingly, though I was well aware that my limited background in college physics was nothing compared to his years of experience at the forefront. "I think we're pretty sure of that, anyway. Molecules, atoms, electrons and protons. I learned all about that in school and I've heard friends talking about things like quarks and moo-ons. Particles are an established fact, not nonsense."

"You mean *muons*. Yes, of course. I learned about all that, too. I learned the standard model. I even proved a theorem about super-symmetry that particle physicists are always making use of, but that doesn't mean it is an accurate description of the real world. I also learned that Columbus discovered America, but by the time you went to school they knew that it was not quite true. Here, let's try this. You say you took a course in quantum mechanics, right? What did they tell you about the speed and position of a particle?"

"That you can't measure them both at the same time."

"Yes, but there is more to it than that. Physicists will tell you it is not only that the particle has a speed and a position that you do not know, the particle does not even *have* a position or a speed until you measure it."

"Yeah, I suppose I've heard that," I said, beginning to wonder whether all physicists are nuts. "That never quite made sense to me."

"And I'll tell you a simple way to solve the dilemma: get rid of the particles. There are no particles after all, only waves. The particle is just a figment of our imagination, and that is why it only has the properties we know it to have. *This* is what Isaac Newton realized when he reconsidered his theory of *Opticks*: all of this around us is a big wave, rippling and moving under the force of gravity. The gravitational effect may make it look like there are particles, but looks can be deceiving, no?"

To demonstrate this last remark, he held the eraser of his pencil loosely between his fingers and shook it up and down quickly. It sure looked like it was flopping around like a rubber tube instead of a wooden pencil, and I had to laugh.

"But," I said, almost starting to believe him, "if Newton knew all of this, why didn't he tell anyone?"

"He tried! That's what his *fluxions* were, a new mathematical notation for the wave nature of reality itself. If only he'd had your non-standard analysis to make it rigorous people might have seen what he was getting at. But, as it was, they thought it was an inferior attempt at defining derivatives and chose to go with Leibniz instead. But he wasn't only talking about functions and calculus, he was talking about the ultimate description of reality. That's what his alchemy was about."

"Okay, so now we get to it! What does it have to do with alchemy?"

"That's what alchemy *is*! Look, today's scientists wouldn't try to turn lead into gold. For them, particles are particles, unchanging by their very nature. But, suppose instead that lead and gold are different ripple patterns in a wave. With the right sort of nudging, you can change one into the other. I'm not just talking about a theoretical possibility. I know exactly what I need to do now, and I'm about to do it!"

"You mean, you're *really* trying to turn lead into gold?!? Come on, even if you can do it, isn't there something more useful you could be doing with this new science of yours?"

"Sure, sure. If today's experiment works, I'll move on to something bigger: producing the *philosopher's stone*. The ultimate, universal substance, entirely unknown to modern science, a quantum superposition of the resonance patterns of all of the other elements. Newton proved its existence mathematically but was never able to work out how to make it. All of the alchemists of his day made this their highest goal. The associated Riemann-Hilbert problem might be so hard that even you would have trouble solving it. . . but that's for another day. Today, we need to turn lead into gold because *that* is what it takes to get the attention of the media."

"Well, there I guess I can agree with you, Professor Stein. If today's experiment works and you turn lead into gold, you could have a TV news crew here tomorrow and. . ."

"But, they are already here! And, today's experiment *will* work. How can you have any doubt? You solved the latest jump problem, no? Well, then we're ready to go!"

It was only then that I noticed the light coming from the living room. I had mistakenly believed that it was sunlight, but this time of year it is not so bright so early in the morning. When I followed the professor through the doorway and my eyes became accustomed to the light, I saw several reporters and camera crews set up there. Dr. Stein went over to a computer terminal in the corner and began typing in the results of my night's computations.

The tall man with the huge pile of light brown hair and unbearably white teeth started talking as soon as we entered. "This is Tom Cannon reporting from the home of Dr. Franklin Stein, a U of M mathematics professor who is either a genius or a certifiable kook depending on who you want to believe. We are here today because Dr. Stein claims to be able to turn lead into gold using an old science called 'alchemy'..."

Meanwhile, the Asian woman in the short skirt was saying "The mathematics department has refused to comment, but Peter Watkins, author of last year's best selling 'The Alien Abduction Diet Plan' insists that Stein's approach is well grounded and sure to produce spectacular results..."

And, most distressingly, the local access channel's young volunteer reporters were doing a terrible job of discussing me and my role in today's spectacular demonstration: "According to Stein, Igor Stravinsky – no relation to the famous violinist – played a key role in his research by solving Rhymon-Dilbert problems, math questions about integrals like those you might have seen in your calculus class..."

While I tried to hide in the corner, all eyes (and cameras) were on Dr. Stein.

"Thank you all for coming here today to witness a tremendous advance in the human understanding of the universe." At this point, he switched on one of those hokey electric spark machines that you see in old horror movies. It was sitting on the coffee table and began buzzing and zapping as the spark climbed higher and higher before returning suddenly to the bottom. "This journey began long, long ago. Al chemie, like al gebra, and even (ironically) al cohol, all owe their origins to that period in the history of Arabia when their scientists were at the forefront of discovery. However, they did not know enough about waves to take it through to its conclusion. This, like so many of their great discoveries was lost to time. Similarly, Isaac Newton who finally knew enough to put together the main ideas did not know enough about his own invention, the infinitesimal calculus, to..."

"Hey Doc," called out one of the cameramen, demonstrating that despite their mock sincerity, the news crews did not have any respect for Stein as a man of science. "We've only got a little bit of tape left here. Can you get on with it and turn the lead into gold already so we can get out of here?"

A few others snickered, whether at the idea that we might actually see lead turned into gold or at the cameraman's rude behavior I don't know, but Stein complied politely, and dramatically. He picked up a grey, metallic bar from the coffee table and dropped it. It shattered the table top and fell to the floor with a loud *thunk*. "This is lead," he said matter-of-factly. "This is a generator of vibrations in the electro-magnetic field," he continued, indicating the spark-making thing. "I place the lead bar on this vibrating platform which, like the field generator, is controlled by a digital computer."

"What kind of computer is it, doc?"

"The kind does not matter. The point is that it vibrates the bar and the electromagnetic field according to the real and imaginary parts of Igor's solution of the Riemann-Hilbert problem respectively so that..."

Though I was already in the corner, behind the recliner, I tried even harder to hide at the mention of my name and hoped that Becka was not watching TV this morning. Then the bar began to shake very quickly, and the pattern of sparks became much more erratic. Space itself seemed to bend visibly. This ripple of space began at the bar but seemed to spread out in all directions, even passing through me—a very strange sensation—before disappearing through the walls.

I was impressed! Dr. Stein seemed to know what he was doing after all. Or so I thought until I looked at him. He looked quite puzzled and concerned. Soon, however, he looked up again at the camera crews and smiled, proudly displaying a shining bar of gold.

The reporters were temporarily speechless. This was not going to be the humorous "human interest" story they had expected. Everyone in the room, even Stein, was only beginning to realize the implications of what we had seen. But, this sense of triumph did not last long.

Beginning with a few drops, soon a steady stream of water was dripping down on us from the chandelier. We all looked up at it, still and silent, noticing the cracks forming in the ceiling, before taking the necessary action. Everyone ran out onto the street as fast as we could. A few seconds later we heard the ceiling crash in and saw the living room flooding through the windows.

Up and down the street, every house was going through the same strange set of circumstances. Wet people ran out into the street as their homes and possessions were soaked in a flood of water.

"I don't understand..." Stein whimpered. He looked miserable. But I was in a pretty good mood: I was actually part of something noteworthy.

"The pipes, Dr. Stein. These old buildings had lead pipes," I explained. "You've turned them all to gold, which may earn the homeowners a nice profit in the long run, but gold is not strong enough to..."

"Of course, I know that!" he snapped. "But only the bar in our experiment should have been affected. The wave should have dissipated before ever reaching... unless..."

His expression vacillated between horror and anger as he grabbed me by the shoulders. "What boundary conditions did you use, Igor? The solution was compactly supported, wasn't it?"

"Oh, well... that didn't seem to be working out nicely," I admitted, "so I used *periodic* boundary conditions this time. Was that, um, bad?"

Behind him I could see an angry mob heading in our direction, accompanied by the local access film crew. They did not seem grateful for our gift of golden plumbing.

"It depends," he asked through a forced smile. "What do you think the effect will be on the world, on technology, on *the economy* now that every atom of lead on the planet has suddenly become gold?"

Ellen Maddow

Three scenes from
Delicious Rivers

Characters

Irma Lurchman—an artist
Lorraine Stone—a single mother
Sy Turner—a school counselor
Joe Hertz—a post office worker
Bill Schmertz—a post office worker
Lily Trillium—a post office worker
Donald Arnold—a post office worker
Three musicians—bass voice, bass fiddle, bass trombone

Production Note:

Ellen Maddow developed Delicious Rivers, a music theater piece on and about Penrose tilings, over the course of the BIRS/Banff workshop series, in collaboration with Marjorie Wikler Senechal. Its first public reading was held at Banff in 2006. The play was subsequently performed in New York and Northampton, Massachusetts.

Scene 3 (Three Fold Rotation)

Lily Trillium stands stiffly, framed by window 2 at the post office. She has tight skin, wide open eyes, and speaks in an intense monotone. Bass Fiddle is in Window 1. He plays as she speaks. The music is mellow, melodious, and circular.

Katrina Ylimaki as Lily Trillium. *Delicious Rivers,* Smith College, October 2006. Photograph by Jon Crispin.

LILY

They say a strong wind's coming in from the West
They say after midnight, or maybe tomorrow
A big, strong wind. Everything will be changing
A storm from the West, a change in the air
Maybe tonight, a big wind, a strong storm.
Clear things out, freshen things up.
Right now, they say, we've got an inversion
A perpetual, barometric depression
My apartment is damp, twenty-eighth floor.
My mattress is thin, my furniture's German
My windows are large but I can't get them open.
If I stand in the corner, I can just see the river.
(This can't go on)

On August 1 at 4:15
A flock of small birds flew in from the East
I felt a slight breeze, I went out of my house
I buttoned my sweater, I walked to the river.
The pier is long, the water is greasy
The fishermen keep what they catch in a bucket
One showed me a mackerel that he had just landed
He couldn't eat it, against his religion
He gave it to me in a brown paper bag
It was a gift, I couldn't refuse it
When I got it home it was still alive.
Its eyes were yellow, its scales were slimy
(This can't go on)

I filled the bathtub and dumped him in it.
I fed him liver, his gills were pulsing.
I've named him Anton. He's becoming enormous.
I can't take a bath, the pressure is building.
At night I dream that he is my lover
In the morning I wash at the sink in the kitchen
In the afternoon I plot his murder.
In the evening I feed him and freshen his water.
Maybe I'll purchase a large duffel bag
And get the doorman to call me a taxi
Leave the bag in the middle of Grand Central Station
And let Anton gasp his last gasp among strangers
(This can't go on)

On the first of August, a flock of small birds
I walked to the river . He gave me a mackerel
I couldn't refuse it. It was soulful and slimy,
A bottom feeder. This can't go on.

He lives in the bathtub. His name is Anton
I'll buy a sledge hammer and crack his scull open

Cook mackerel stew with dill weed and onions,
And eat him for dinner, this can't go on.

My skull is on fire, I stand in the corner.
A perpetual barometric depression.
Everything will be changing, a storm from the West.
They say after midnight, this can't go on.

The pressure is building
I dream he's my lover
I'll call Rabbi Gutbaum
To give us his blessing
Then we'll swim out to sea
Away from the weather.

A big storm is brewing
My windows are rattling
I open the door
The room fills with water
His blood oozes out and
Drains into the sewer.

I rip off my sweater
I walk to the river
The water is greasy
The fisherman's sneering
The fish was a gift
I couldn't refuse it.

They say after midnight his eyes are yellow
He fills the bathtub, soulful and slimy
His gills are pulsing, twenty-eighth floor
I plot his murder, my skull is on fire.
He's still alive
He's still... he's still...
He's still...

He is
Still.

Excerpt from Scene 4 (Penrose Patterns)

Irma Lurchman, Sy Turner, and Bass Trombone are sitting on a bench in front of the post office. Two dogs are tied by their leashes to the ends of the bench looking anxiously toward the post office doors through which their owners have disappeared. Lorraine Stone arrives with a shopping cart full of bags of old clothes.

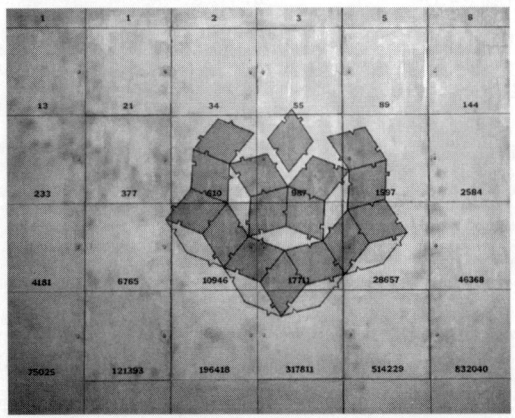

Penrose tiles (animation by Anna Kiraly). *Delicious Rivers*, Smith College, October 2006. Photograph by Jon Crispin.

LORRAINE
What's the matter with him?

IRMA
Oh, it was his birthday.

LORRAINE
Happy birthday, Sy. Did they throw you a party?

SY
They could have, but they didn't.

LORRAINE
So instead you...

SY
Instead I took my family on a trip.

LORRAINE
Nice! Where'd you go? Caribbean? Europe, Fiji?

SY
I could have but I didn't.

IRMA
So instead you...

SY
Instead I went to Florida.

LORRAINE
Florida?

SY
I used to spend my summers in a cottage that my parents rented at a lake—Mock Lake, it was called. The weeds around the edges had a wonderful, an evocative smell. The water was clear and tepid, the afternoon light—you know that late afternoon light toward the end of summer when the breeze picks up and every blade of grass, every tiny bug, every rock on the bank of the lake has an enormous shadow and time stretches out forever and your bare feet are licked by the warm tongue of your old dog and...

IRMA
And?

SY
I would swim out to the center of the lake and lie on my back with my arms stretched out. The world was an enormous pinwheel and I was the pin at the center.

TROMBONE
Music. (Everytime there is music, the actors shift positions.)

IRMA
So you wanted to experience it again?

SY
I wanted to show it to my family, but the cottage was gone, burned to the ground in 1969.

LORRAINE
So instead you...

SY
So instead we checked into the Mock Lake Marriott. We went down to the lake, but the lake was dried up—cracked gray mud as far as the eye could see.

IRMA
Too bad! I bet you felt terrible.

SY
I could have but I didn't.

LORRAINE
So instead you...

SY
Instead I tried to brush it off. I tried to forget all about it.

IRMA
You tried to...

SY
I tried to but I couldn't.

IRMA
So instead you...

SY
Instead I snuck off in the late afternoon, I put on my bright green jogging shorts and set off for a run by the dried-up lake.

TROMBONE
Music. (Shift.)

SY
I got to the shore, the weeds were still there,
Pungent as ever, the light was just right.
I could feel the Earth tipping, The pinwheel clouds.
And then I turned, and I turned again.

LORRAINE
You turned and went back to the Marriott?

SY
I could have, but I didn't.

IRMA
So instead you...

SY
Instead I stepped off of the bank.
I thought I'd walk out on the cracked gray mud
Out to the center of dried-up Mock Lake.
And look back at the weeds and the light on the rocks,
And the spiraling sky.
I thought that It might be the same,
The same but not the same.

LORRAINE
It must have been strange.

SY
It was worse than strange
The light was just right, but I started to sink

And then I was stuck and sinking fast,
Up to my thighs in stinky mud.

LORRAINE
Did you call for help?

SY
I could've, but I didn't.

LORRAINE
So instead you...

SY
Instead I looked into the distance.
I saw a man jogging, but so far away,
So tiny and small, I thought to myself.
"I am going to die."

TROMBONE
Music. (Shift.)

LORRAINE
But here you are, you didn't die.

SY
I could've, but I didn't.

IRMA
So instead you...

SY
Instead I started to wiggle and flop
and push and roll and cling and slither
and twist, until I was free.
I wiggled away, like a seal in the sand,
One shoe lost in the muck of Mock Lake.

LORRAINE

I bet you felt lucky.

SY

I could've but I didn't.

IRMA

So instead you...

SY

Instead I was ashamed.
I limped along, to the Marriott
I dripped across the lobby floor,
Out to the patio, no one was there.
I dove into the pool and sank to the bottom
Until I was clean.

TROMBONE

Music. (Shift.)

SY

I'll never try to go back there again.

LORRAINE

Well you could.

SY

I could, but I won't.

Excerpt from Scene 7 (Mr. Schmertz and Mr. Hertz—Heterochiral Reflection)

The bench outside the post office. Trombone and bass fiddle are playing. The music is syncopated and percussive. Joe Hertz and Bill Schmertz do a heterochiral dance. They are snappy movers —like a left and right hand, separate but in touch.

Left: James Kules as Joe Herz; right, Ezra LeBank as Bill Schmerz. *Delicious Rivers*, Smith College, October 2006. Photograph by Jon Crispin.

BILL

Fishing on your day off again, Mr. Hertz?

JOE

Yep, end of pier 40, gray water, gray sky, bracing wind.

BILL

As usual.

JOE

Nothing like it. And you, Mr. Schmertz?

BILL

Satellite dish, Discovery Channel-collecting facts.

JOE

As usual.

BILL
Facts don't lie, Mr. Hertz, Am I right?

JOE
I keep what I catch in a bucket, Mr. Schmertz.

BILL
Got those facts filed away on a shelf in my brain, Mr. Hertz. Entomology, Astronomy, Mathematics. The life of the bee and the death of a star.

JOE
Scale, gut, filet, wrap it in plastic and file it in the freezer. I eat fresh fish every night of the week. You are what you eat, Mr. Schmertz.

BILL
I am what I know, Mr. Hertz. Concrete facts, you can always trust them. Am I right? People are weak—feckless and shifty. With facts you always know where you stand.

JOE
Concrete pillars wedged in rock, under the water. I stand on the pier, I breathe the air, I know who I am.

Music and dancing.

BILL
When I was younger, Mr. Hertz, I was a mystery to myself—round and red and noisy, in a family that was dry and whisper-thin. Yes sir, a big red dog in a pile of dry leaves.

JOE
When I was younger, Mr. Schmertz, I was obsessed with the mysteries of physics. I spent long hours, alone in a room, my skull filled with formulas that burned like fire in my mind's eye. Physics was a fierce and distant God that might someday whisper the truth to me.

BILL
One day they told me the truth: I was adopted. They say they found me under a truck, three years old, I'd witnessed a murder. They said I'd seen too much, maybe I had, I don't remember. They say they saved my life. What were the facts? I couldn't remember, Mr. Hertz.

JOE
One day after hours of reading a book on entanglement theory, I looked up and the whole world had become a mathematical model. I couldn't remember what was real. I walked out of my room and into the street—everything I saw was in code. My life was a graph. My breath was a series of points on that graph... I was afraid, Mr. Schmertz.

BILL
I thought I could know the truth if I could only find my mother. I finally did, she was round and red and tough. We met in a bar in a Polish city. She held my hand, I looked at the table. We had nothing to say to each other. I was afraid.

JOE
I found I was walking barefoot on hot blacktop. I felt my feet burning, I came to my senses. But I never went back to that room, to those books. Now I fish to feel alive, Mr. Schmertz. Whenever I'm free, I go fishing.

Music and dancing.

Marjorie Wikler Senechal

The Last Second Wrangler

"No, I never met him, not in person," Michael Longuet-Higgins replied. "But in high school I proved a theorem about the Simson line, what happens when the point moves around the circle.[1] My teacher was very excited and sent it to an expert, in this case Neville."

We were chatting at a reception at the Institute for Advanced Study celebrating the 2006 publication of Siobhan Roberts's *King of Infinite Space: The Man Who Saved Geometry*, a biography of H. S. M. Coxeter. Coxeter, a British mathematician transplanted in Toronto, died in 2003 at the age of 96. Eric Neville had been his mentor and friend; Coxeter had been mine, and Longuet-Higgins's.

The eminent Professor Neville had doused young Michael's excitement. "He replied that the result had been proved by Steiner in the 19th century and I should stay away from this kind of geometry, that it was a dead end. I thought, oh well, I'd enjoyed doing it," Longuet-Higgins said. But, as such things so often turn out, it wasn't a dead end at all. "Years later I was studying the propagation of sound underwater. It goes very far very fast. A disturbance in the Indian Ocean can be heard halfway round the world. My result on the Simson line helped me understand why. But Neville had no interest in such applications."

"I met him once," Freeman Dyson joined in. I'd interrupted a conversation between these renowned mathematician/physicist friends. "I was a

---

[1] Draw the circle defined by the vertices of any triangle. Let P be any other point on the circle. Mark the intersections of perpendicular lines through P to the triangle's edges (extended beyond the circle, if necessary). These three points always lie on a line (named for Robert Simson, 1687–1768, though there is no proof that he proved this theorem). For more details, see http://asa.aip.org/vol13no4.pdf/.

student at Cambridge during the war. There were very few of us in the math club."

"You were the president, weren't you?" Longuet-Higgins asked. Dyson nodded.

"There weren't many mathematicians around, so I invited Neville to come from Reading to give a talk. He took the train; I met him at the station. I remember only that he was very boring. I was very happy to put him back on the train. Why do you ask?"

I ask because I'm writing a book about Dorothy Wrinch, a colorful, brilliant, and controversial character; and, through her, about crystals and symmetry.[2] She held doctorates in mathematics from London and Oxford and wrote dozens of mathematical papers, but she's not remembered for them. You'll find her instead in histories of genetics and chemistry, albeit in caricature. Wrinch catalyzed the science of proteins in the 1930s with the first geometrical model for protein architecture. The catalysis was quick, the temperature fierce. Embattled, then sidelined, she defended herself to the end. And, in the end, though she was wrong about proteins, she wasn't as wrong as her enemies said.

Dorothy Wrinch has been called many things: a genius, an interloper, a fanatic, an opportunist, a prototype for Rosalind Franklin. I knew her well in the last decade of her life and saw flashes of all of those facets. But, first and last, she was a mathematician in spite of herself, and Eric Neville was her truest friend.

$$\infty$$

He was "tall, loosely built though not athletic; with iron-grey hair... He wore rimless spectacles," said Walter Langford. Not to me—Langford died before I began writing my book—but to friends in Reading in 1978 gathered to celebrate his donation of Neville's books to the university library. Behind those glasses, Langford continued, "the eyes were often glowing with the light of a puckish humour." Though not on the day Dyson met him at the train.

Langford prepared Neville's papers for the Reading archives; they fill only three or four boxes. If Neville saved personal letters, he saved them somewhere else, or his brother Raymond took them. I was more surprised to find very little professional correspondence in those boxes; Neville must

---

[2] "'The Last Second Wrangler" draws on my forthcoming book, *I Died for Beauty*.

have received, and written, hundreds of letters in his long years of service for committees and professional societies, but none remain at Reading.

His obituaries recount the milestones of his life. Eric Harold Neville was born in London in 1889. His high school teachers noted his unusual mathematical ability and prepared him for Cambridge; he entered Trinity College in 1907. Two years later, Eric placed second in the six-day marathon called the Mathematical Tripos, thereby earning a permanent niche as Second Wrangler in the mathematical pantheon.[3] He spent most of his career at the University of Reading. Well known and greatly respected in the wider British mathematical community, he served stints as President of the Mathematical Association and Chairman of the British Association's Committee on Mathematical Tables. Neville married Alice Farnfield in 1913; their only child, a son, died in infancy. "Mrs. Neville was a very gracious lady, with a strong social sense," wrote Langford. "She was the complete foil to the idiosyncrasies of E. H.'s genius. As host and hostess, they were unsurpassed in my experience and many others share this opinion; the annual student parties at their lovely house on Castle Hill, Reading, were renowned in my day." Alice Neville died in 1956 and "those of us who knew him closely have the feeling that he never reconciled himself fully to her passing." Neville died in 1961.

But, admitted Langford, E. H. was "by no means an easy person to know or to understand really well."

∞

The Mathematical Tripos was the centerpiece of a Cambridge education for two hundred years. In its heyday, from the late Victorian era to 1910, the top three, or four, or five, Wranglers were instant celebrities, feted like sports heroes, lionized by the press, photographed for postcards, held in awe all their lives. They trained like Olympic contenders, with private coaches. Townspeople bet on them.

"The English have always had more faith in competitive examinations than any other people (except perhaps the Imperial Chinese)," said C. P.

---

[3] Cambridge students winning First Class Honours on the Mathematical Tripos are still called Wranglers; the term dates to the days when the Tripos was an oral exam, in Latin. From 1753, when the Tripos began, to 1909, the Wranglers were ranked by their scores, the top scorer being Senior Wrangler, the next Second Wrangler, and so on.

Snow.[4] The articles of faith in the Tripos were two: first, that mathematics was "the ideal foundation . . . to educate the future leaders of Britain and her Empire," and second, that this grueling test of mental agility measured the qualities needed to shoulder this burden. And so, of course, it proved: the long list of Senior and Second Wranglers includes four Lord Justices of Appeals, four prominent actuaries, two Bishops, an Archdeacon, a president of the General Medical Council (who spoke 18 foreign languages fluently), and a Deputy Speaker of the House of Commons. But also famous physicists, among them Michael Faraday, James Joule, James Clerk Maxwell, William Thompson (later Lord Kelvin), George Stokes, John William Strutt (later Lord Rayleigh), George Green, Peter Guthrie Tait, and J. J. Thompson.[5]

Many famous physicists, but few world-class mathematicians. That's because, Godfrey Harold Hardy (Fourth Wrangler, 1898, and one of those few) insisted on every possible occasion, the Mathematical Tripos didn't measure aptitude for creative work in mathematics. No exam could. Abolish the Tripos! he thundered. The competition for rankings and the emphasis on clever tricks had frozen the curriculum and retarded the development of British mathematics for a hundred and fifty years. In 1908 Hardy won a partial victory: in 1910 and forever after, winners would be listed alphabetically, and all bets would be off.

Young Eric Neville was deep in preparation for the 1910 gold crown at the time of this shattering announcement. Loathe to lose his chance at immortality, he took the exam a year early. The stakes were high. Not only was the 1909 Mathematical Tripos the last to be ranked, it might settle an ancient rivalry. Over the years the Tripos had become a contest between Trinity and St. John's next door: going into this final round, these two colleges boasted 55 Senior Wranglers each.

"Who will be the last senior Wrangler at Cambridge?" one headline brayed. In the photograph Neville, one of six leading contenders, looks much as Langford described him, but younger than his twenty years, and altogether humorless. His black hair is slicked back, his glasses are rimless, his face thin. He saved the news clippings all of his life.

---

[4] C. P. Snow, Variety of Men, New York, Scribners, 1966.
[5] D. O. Forfar, "What became of the Senior Wranglers," Mathematical Spectrum, Vol. 29, No. 1, 1996.

Cambridge, Tuesday, May 25

The results of the great mathematical examinations which began on Monday, May 17, were announced to-day, the most important, those looked forward to with the keenest expectancy, being:

Senior Wrangler: Mr. P. J. Daniell, Trinity

Second Wrangler: Mr. E. H. Neville, Trinity

Third Wrangler: Mr. L. J. Mordell (of Philadelphia, U.S.), St. John's

Great excitement prevailed among the largest crowd ever assembled in the Senate House on Tripos mornings. . . . The striking of the hour when results are announced was awaited with ill-concealed impatience, the examiners meanwhile standing unmoved in the gallery with the fateful list ready to hand. In absolute silence the clock chimed the hour, and Mr. M. Dodds announced Mr. Daniell to be the last of the Senior Wranglers. Trinity men shouted frantically and cheers were renewed when it was known that another member of the college was placed second. The St. John's section retaliated at the mention of the third Wrangler.

"I blotted my copy-book and was only Third Wrangler," Mordell lamented in 1969. "I think I could have done better." [6]

Daniell, Neville, and Mordell: they're adjectives now, in the mathematical custom: the Daniell integral, Neville Theta Functions, Mordell's conjecture (its role in the solution of Fermat's Last Theorem was posthumous). Mordell is the best known of the three. I met him a few times at the University of Arizona when he visited in the early 1960s. He was in his seventies then; I remember he wore thick glasses. I don't remember an American accent; he seemed the quintessential English don. I was too awed to ask him about mathematics or anything else.

---

[6] L. J. Mordell, "Reflections of an Octagenarian Mathematician," lecture to the Philadelphia Section of the Mathematical Association of America at Swarthmore College, November 22, 1969.

∞

Neville went on to win Cambridge's coveted Smith's prize in 1911 and a Trinity Fellowship, the most prestigious in England. His fellow Fellows, the cream of the mathematical crop, included G. H. Hardy, J. E. Littlewood (Senior Wrangler, 1906), and G. N. Watson (Senior Wrangler, 1907). In 1914, Neville traveled to India to lecture at the University of Madras; there he met the young genius Srinivasa Ramanujan and persuaded him to come to Cambridge to work with Hardy and the others. (Ramanujan's story is one of the most remarkable, and tragic, in the history of mathematics; see Robert Kanigel's biography, *The Man Who Knew Infinity*.) Then war broke out, and that changed everything.

Neville's examiner for the Trinity Fellowship was Bertrand Russell, a lecturer at Trinity, famous as the co-author of three massive tomes immodestly titled *Principia Mathematica*.[7] Despite, or rather, because of its austerity, *Principia* blew a fierce wind through mathematics, stripping away the debris of history: cluttered notation, labyrinthine proofs, encrusted applications. Just as—and just when—Picasso painted helter-skelter cubes, Schoenberg revised musical tones, and Planck discovered the elusive quantum, Russell declared symbols, and relations between them, to be mathematics' true essence. Geometrical language "is only a help to the imagination," he wrote. "Mathematics may be defined as the subject in which we never know what we are talking about, nor whether what we are saying is true." From Russell, said Langford, Neville "learned to distrust the foundations on which Cambridge mathematics was then based, and to sharpen his own formidable logical equipment."

Russell's influence on twentieth-century philosophy and mathematics and, later, on social thought is well known. Nor need I discuss Russell's theory, and practice, of free love, or his problematic relations with his many wives and lovers. Neville and Wrinch moved in another orbit around Russell, somewhere between his unfortunate intimates and adoring vast public: the orbit of devoted acolytes. Were they burned by his heat, stung by his cold? Which waves of his broad energy spectrum did they absorb? Neither Neville nor Wrinch modeled their careers on his: neither became a logician, or a philosopher, or a public intellectual. But they bore his stamp for the rest of their lives.

---

[7] *Principia Mathematica* was the title of Newton's masterpiece.

I see it in a review of Neville's forbidding *Prolegomena to Analytical Geometry in Anisotropic Euclidean Space of Three Dimensions* (1923):

> To barbarians it will seem to cut right across the course of modern geometry with an independence which shows itself in nomenclature and notation, in absence of references, and most of all in the limitations which the author has placed upon himself in the selection of his material. This is partly accounted for by the fact that Prof. Neville is avowedly a disciple of Mr. Russell, whose well-known aphorisms are scattered over the book.

Wrinch wrote that way too. Idiosyncratic nomenclature, nonstandard notation, absence of references, limitation of selection—these are complaints her critics voiced before and after she plunged into proteins. They're mine as I struggle with her papers and books.

When Russell served prison time in World War I for his pacifist activism, Wrinch ran errands for him and supplied him with books. Neville worked in a London hospital throughout the war years. "Although his eyesight would have barred him from any combatant service," Langford said, "his whole nature revolted at the thought that, in any circumstances, man should kill his fellows, and he declared that as his conviction from the beginning. In a letter which I had from Lord Russell recently, he writes about Neville and comments that this 'was an obstacle, throughout his life, to academic advancement in spite of his brilliance, as many educational authorities think that a man is not fit to teach the young if he dislikes having them slaughtered. I was fond of him, particularly because of his gentleness.'"

Neville maintained his principled pacifism in World War II. Russell and Wrinch supported the fight against Hitler. "If you are able to hang on to what you thought in the last war you are lucky so far as your inner Geist is concerned," she wrote to Neville in 1940, "but I know so unlucky in hosts of other ways, for of all things having min. opinions is the worst faite (sic) I think. One misses the comfort of common feelings with others for which I always long." In the scientific community she would soon be a minority of one.

"One can only conjecture," Langford concluded, "whether his Fellowship might have been renewed had he not declared his opinion, but the fact is that it was not and he moved from Cambridge never to return."

Reading is an attractive city on gently rolling hills in the agricultural countryside, halfway between London and Oxford. Entering the campus, I asked a student for directions to the library; she walked with me. It's boring here, she said. Nothing ever happens. Not so, says the Reading Station website. In 1919 "T.E. Lawrence (Lawrence of Arabia) lost the 250,000-word first draft of his *Seven Pillars of Wisdom* when he misplaced his briefcase while changing trains." Neville joined the faculty of the University Extension College[8] that same year. When, seven years later, the college received a royal charter and became the University of Reading, he was its only mathematics professor.

$$\infty$$

Heart's love; dearest Erice. He wrote on onionskin with a fine-nibbed pen, cramming long sheets with love—and advice, mathematical calculations, updates on his book manuscript, on his poor health, and on the tribulations of war-time Britain. She typed her replies and saved carbon copies. She worried about his health, complained about her opponents, thanked him for helping her. Though Baltimore was far from the front, her little daughter Pamela still dreamed of bombs and Nazis. She never wrote of love, out of discretion—or maybe honesty. Some letters crossed, some were lost, but those we have sketch another Neville, and his parallel life. Their letters span twenty years, their relationship thirty.

They met in the war years at a rally for Russell. Or perhaps they met in Cambridge before the war. The mathematical world was finite then: everyone knew everyone, and Wrinch knew them all. A student at Girton, an unranked Wrangler in 1916, she studied with Russell in London, began research under Hardy, and stepped in to teach for Watson at University College when he moved to Birmingham.

Or maybe they met in the 1920s, after Wrinch moved to Oxford and began teaching in the women's colleges. For years they served together on the Committee on Mathematical Tables, which oversaw the computations of enthusiasts working at home. In those days before computers, tables of values of logarithms, sines and cosines, Bessel functions, the Gamma function, all the special functions of mathematics and mathematical physics, were calculated and proofread by hand. Neville "had always been strangely

---

[8] His application for a professorship at the University of Otago, in New Zealand, was, evidently, unsuccessful, even though Hardy supported it.

attracted to tabulation processes;" perhaps Wrinch was too, or perhaps she joined because her husband, John Nicholson, was the committee's chairman.

In 1930 Nicholson was hospitalized briefly for alcoholism, then permanently for lunacy. Wrinch obtained a legal separation. Soon afterwards, and not coincidentally, she retooled herself as a mathematical biologist. Today Google finds 778,000 sites for "mathematical biology"; back then the very idea was unheard of, unnamed, and unwelcome. But Wrinch knew what she was doing or, rather, what she hoped to do. She'd studied the shapes of living things—birds' wings, sponges—for many years, and she'd written on mathematical physics. Now she would bring her interest and expertise to bear on the shapes of cells, chromosomes, and proteins.

Neville set his own work aside to give her a hand. In her first long paper after reincarnation, Wrinch thanked fourteen colleagues, Neville the only mathematician among them. He appears in footnotes and captions again and again: "The author offers her thanks to E. H. Neville for his advice and criticism"; "Reproduced by kind permission of Prof. E. H. Neville, University of Reading, England, who kindly constructed this model."

The model in question was a "cyclol", one of a series of molecular cages that Wrinch believed captured the geometry, if not the chemistry, of proteins. She, and her model, provoked thought—and quarrels. The brightest stars of science got into the fray: Irving Langmuir, Niels Bohr, Harold Urey on her side; Linus Pauling, Dorothy Crowfoot Hodgkin, W. L. Bragg, and J. D. Bernal on the other.

The quarrel got nasty. She didn't know any chemistry! Who *really* discovered those cages, and when? Not I, said Neville in a letter for showing about: "Dear Dr. Wrinch, It must have been in the summer of 1936 that I first made the models of the cage structures which you were proposing for the proteins... I hope these recollections are accurate enough for your purpose. With kindest regards, Yours ever, Eric H. Neville."

Wrinch and Langmuir chose insulin as a test case. The test was arduous: first crystallize the protein, then bombard it with x-rays, then search the pattern of diffraction spots for clues to its structure. Dorothy Hodgkin had crystallized insulin and x-rayed the crystals, a major achievement, but the pattern of spots defied interpretation. Wrinch and her supporters claimed the spots showed a cyclol structure; Hodgkin and hers insisted they did not.

How could anyone tell? In fact, they couldn't. "No one understood anything about proteins back then; that's why they fought so viciously," Caroline MacGillavry, a crystallographer and Wrinch's friend, told me forty years later. It was a battle of the deaf. "My colleague Neville, whom you met here last September and at Nottingham, has had several conversations with Bernal, on my behalf," Wrinch wrote to Langmuir, "but he gets the same impression ..."

The crystallographers were trying to deduce atomic positions from contour maps called Patterson diagrams, maps derived from x-ray data. Wrinch made a breakthrough.[9]

First, forget the contours, she said. Look just at the points at their centers, the peaks.

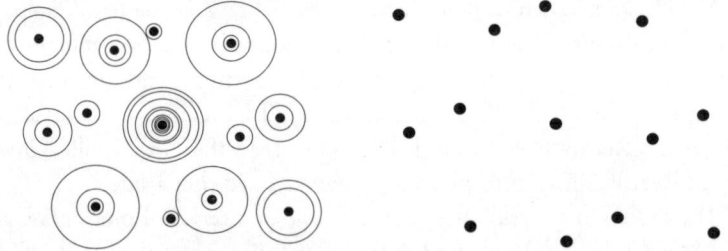

Just the *points?* they gasped. Yes, just the points, she snapped. The peaks show the distances between the atoms in the crystal. Points do that just as well.

The distance vectors radiate from the center like a star.

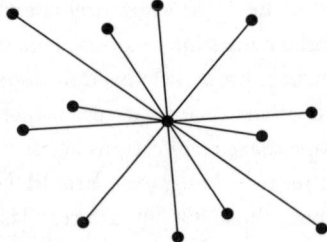

It's a geometry problem, she said. Given the star, find the relative positions of the atoms.

---

[9] D. M. Wrinch, "The Geometry of Discrete Vector Maps", *Philosophical Magazine*, Ser. 7, Vol. xxvii, January 1939.

Just as the night sky is a cultural Rorschach test (Western eyes see hunters and dippers; Chinese eyes see dragons and birds), the same set of points can be joined as a star (above) or as a pinwheel (below).

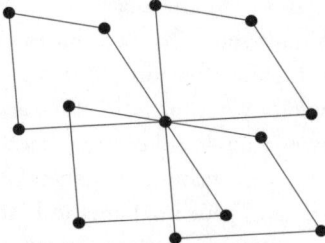

The arms of the pinwheel, all of them congruent, are pinned at the center point. Take any arm and connect its corners to make a kite:

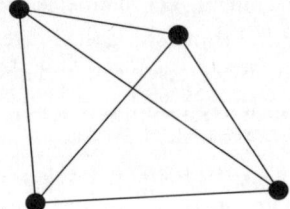

Look at the star again: all its vectors show up in each kite as edges and diagonals. Since each kite is a truss-work of inter-atomic vectors, its corners are the atomic positions we are looking for. Wrinch gave a step-by-step algorithm for finding the kites in the stars.

Bernal didn't get it. The battle raged on.

Just once, at the height of the quarrel, Neville stepped out from the shadows. "Dr. Wrinch is undoubtedly right," he wrote in *Nature*. "The Patterson diagrams contain far more information than was suspected before she began to study the published insulin diagrams for herself. In fact, reconstruction of a discrete point-set from its vector map is a systematic process."

She *was* right about the Pattersons, but few noticed at the time, because she was so wrong about insulin.

∞

Dorothy divorced John in 1938. Eric did not divorce Alice.

"You spoke to me in a time of great stress saying that if there were an outbreak of war you would be going to America," wrote Lynda Grier, principal of Lady Margaret Hall. Though she'd just been appointed a Somerville Fellow, Wrinch booked passage for herself and Pamela on the U.S.S. Washington as Hitler's army crossed into Poland.

From across the Atlantic, in the darkness of wartime, Eric poured out his love. "Darling, I want you so badly. Lying in this bed in a drab room in Bath, in the middle of a dull afternoon, with a picture of you on the table, suddenly the longing for you took hold of me and racked me. Dorothy love, my body is in pain and my mind is numb because you aren't here."

He proffered advice. Don't write joint papers, people will play down your role in them. Don't lend your mathematical talents to the war effort, biochemists won't take you seriously. If a good job comes your way, forget about Oxford...

"Long letter from Margery today," he wrote. Margery Fry, the former principal of Somerville College, was Pamela's "godless godmother" and Wrinch's very close friend. "I think she is the bravest woman I've had anything to do with, for she owns to terror, and she's no official duties to compel her, but she goes back to London to give a helping hand when she has plenty of perfectly good reasons to stay away."

"I think GHH is a beast in his new book about Ramanujan," wrote Dorothy, "if judging by Mordells review he hasn't given your story with R."

"The great tragedy of Ram.," Eric replied, "is that the war cut him off from people with whom he could have done far more than with G.H.H."

"I talked seriously with Donald Coxeter ten days ago in Toronto," she told him. "They are OK though Rene has not heard from her family in Holland since the invasion. They hold fast to your ideas and seem to have insulated themselves from the world. They worry a little because Rene's refusal to go to a weekly knitting socks for soldiers party of the math faculty has led her to being practically cut by the others but they seem sure they are right."

"I had a letter from Donald at Christmas," he said, "and he said he'd seen you and Pam. And that's all he did say. Why didn't he know that I wanted to know everything else about you—if you looked well, seemed happyish, if your work was going down well, and, and, and etc...this bloody war. It seems more and more impossible that I shall see you till it's over. I told Donald months ago that I would give anything to get away from Reading if I could find a junior post, and he is blankly discouraging."

I will come to you soon. Can you live in poverty? For that's all I can offer you...

But then: "Heart's love, Your distressful letter of Dec. 10 has arrived." (What had she written? She didn't save the carbon.) "I can't see how I can offer you a hearth and home till this ghastly war is over."

Margery to Dorothy: "I don't think he—in any generosity—could ask you to wait for him indefinitely, and I'm old fashioned enough to respect him for not throwing over altogether his responsibility to an unloved wife —it's one of those situations which doesn't seem to have got a 'right' solution."

Dorothy to Eric: "Oh dear, it has been a sad history 1930 - 1939 hasn't it. I hope that the FG [as they referred to Alice] may know and believe how little we ever had if it gives her pleasure, as I expect it would . . . and yet you talk of memories. I don't follow you there very well. Of course you saved my life in other ways so often, sums and cyclols . . and I am for ever grateful for this. Never think me ungrateful please, Erice, for I know you gave your time and thoughts to me very often."

Dorothy married Otto Glaser, a biologist at Amherst College, in August, 1941 and joined faculty of near-by Smith. "Heart's love," Eric wrote, "I have known for a long time that my ghostly love couldn't stand between you and anything warm and human and immediate, but wasn't ready for anything so swift. I am glad you are delirious, and I hope you will be always happy."

∞

Neville glumly soldiered on (may he forgive the metaphor): he served on committees, taught at Reading, lectured to Dyson and other students at Cambridge, finished the book on elliptic functions for which he's still remembered, and sent note after note to the *Mathematical Gazette*. He "threw himself with great zeal into the work of building up the department. Honours students were inspired by the brilliance of his lectures and the immensity of his erudition," his colleague T. A. Broadbent said, but Neville complained that "the Honours School of Mathematics is too small to be healthy for either students or staff. The problem is insoluble so long as Cambridge and Oxford are in a different education universe from the other universities of the country."

"Looking back in 1943," wrote a university historian, "E. H. Neville could roundly assert that he had only ever had three pupils at Reading who were worth teaching. His memory was slightly at fault. Since 1926 the University had awarded just four Firsts in Mathematics."[10]

"This state of affairs is a lamentable waste of the abilities which Mr. Broadbent, Mr. Goodstein and I have brought to the service of our profession," said Neville.

Neville wrote eight books; six were published in his lifetime[11] and Langford prepared one for posthumous publication. The eighth, a handwritten treatise on differential geometry, has never been typeset. He wrote sixty-nine papers on pure mathematics (geometry, number theory, differential equations, elliptic functions and elliptic integrals) and several about teaching; and then there were all those notes to the *Mathematical Gazette* (I gave up counting them).

He left no masterwork. "His friends may sometimes have wished that so brilliant and versatile a talent could have been harnessed to some major mathematical investigation of the time, whereby the bounds of our knowledge might have been materially advanced," wrote Broadbent, but "what he chose to do, he did to perfection and though we may quarrel with his choice, we cannot deny the greatness of his gifts."

∞

They stayed in touch.

Eric to Dorothy, Good Friday, 1950: "I went to London for the op. and had it a fortnight ago, on March 21; I stayed in the nursing home—Fitzroy House, where I visited Ramanujan several times 30 years ago! . . . Your last letter was brought to me yesterday. Your sympathy about mother was precious. I found among her papers snaps of you which she had treasured. She always wanted to know if I had news of you."

His health, never good, grew worse. He retired in 1951.

"It's a long time since I've seen anyone who has seen you and could talk to me about you," he wrote. I don't even know if your hair has begun to lose its colour; I find it quite impossible to think of you with gray hair, and I wonder how you wear it now. Should I really know you half a mile away, or would you have to walk right up to me and tell me who you are?" Otto died, Alice died; their correspondence picked up. They met again, but did not marry. Dorothy immersed herself in a final defense of her he-

---

[10] J. C. Holt, The University of Reading; The First Fifty Years, Reading University Press, 1977.
[11] The Prolegomena and Jacobian Elliptic Functions, two sets of mathematical tables, and two extended essays.

retical protein theories. Eric, her steadfast confidant, adviser, and enabler, found a publisher in Denmark. In 1960, wracked with illness, he saw the first volume through the press. It sank like a stone. She carried on; the second appeared in 1965, four years after his death.

Dorothy to Raymond Neville, (telegram): PLEASE GIVE ERICE DEVOTED LOVE

Raymond Neville to Dorothy, August 20, 1961. "Your telegram arrived just before we left for Reading on Saturday. I gave Eric your message but it would be wrong of me to say that he received it as for the hour we were there he gave no sign of recognizing who was there or what was being said to him. As you know it is only in the last few weeks that he has deteriorated so rapidly. Three weeks ago we got him into the Royal Berkshire Hospital where tests were taken and the diagnosis was anaemia from which I was assured that at his age the recovery would be a miracle... There is I'm sorry to say nothing more to add except that we share a mutual sorrow."

$$\infty$$

"We have been much exercised in our minds, Neville and I," Wrinch told Langmuir in 1938, "about the degree of multiplicity of atomic maps having the same vector map . . . and with the interrelations of such atomic maps."

For, in complicated cases, different pinwheels can match the same star. Then the algorithm forks, each branch leading to a different picture. And all of the pictures are true.

$$\infty$$

Neville's books were dispersed throughout the library. Carol Spiers, a university cataloguer, reassembled the collection a few years ago. In its temporary home in the library subbasement, I saw a third edition of Newton's *Principia*, books by Jacobi, Cauchy, Laplace and other mathematical giants, and many more modern.

In the pages of Jacobi's *Fundamenta Nova Theoriae Functionum Ellipticarum*, Spiers found a letter from Neville's old friend Watson, dated five months before Neville's death. Watson too had once loved Dorothy Wrinch, I'm told, but that was long, long before. The letter is a technical discussion of elliptic functions, until the last lines:

> I imagine that your surmise about being the youngest surviving Wrangler of the period ending 1909 is correct. In your year

Rau (Trin) was 9th (I do not remember him) and Bonhote (Clare) was 10th. He went to Forsyth's lectures on diffns in 1908-9, when Forsyth put me on the job of marking examples. I forget what happened to him subsequently.

A. E. Housman was a Fellow at Trinity in the years just before the Great War. He would have dined at High Table with Neville and the other Wranglers. After dinner, they adjourned to the Senior Common Room for port and walnuts. I imagine them there, ensconced in massive leather armchairs around the fire. Housman is reading his poem, "To an Athlete Dying Young":

Now you will not swell the rout
Of lads that wore their honours out,
Runners whom renown outran
And the name died before the man.

Young Neville and the others applaud him.

Manil Suri

The Tolman Trick

The Wolf River emerges from the Black Forest and winds across the snow-covered land. Its water is so cold it is syrupy, a snowball gets carried several yards before it melts. Chunks of ice floating downstream curve gracefully around bends. Fluid particles move along smoothly, their velocities varying between orderly bounds. But then they encounter an obstacle, and their streamlines begin to break down. The river carries on, even though mathematically, a solution to its flow equations cannot be found.

Which was nonsense, of course, as his mentor György would have said: "Nature always provides a solution, it's we mathematicians who are not smart enough to validate her." Tolman scrubbed the steam off the cab window to get a better look at the river. There was a stub of snow projecting from the bank into the water, probably the remnant of an old stone bridge. He was tempted to ask the driver to stop so that he could take a closer look, watch the water hit the corner of the stone and loop around in its complex patterns. Patterns that engineers had recognized, even calculated for decades, but which the mathematical theory had not been able to substantiate. Not, that is, until two years ago, until the publication of his landmark paper on fluid flow around a corner.

Tolman marveled at how dramatically the paper had changed his standing in the pecking order. For years, he had clawed and scratched with everyone else in the barnyards of academia for scraps of recognition. Suddenly, he was judged worthy of a seat at the table. Suddenly, he could reach out and nosh from the platters of conference invitations and panel appointments he had only been able to smell before. Attendees at conferences sought him out, instead of walking away when they did not recognize the name on his badge. Overnight, he had become a star, surrounded by a personal solar system of students and junior researchers. Already, some papers had started referring to his argument as "The Tolman Trick" because of the cunning way it bypassed all the difficulties that arose in trying to resolve the corner. (Personally, Tolman would have preferred that his colleagues use the more respectable "Theorem," though he had to admit that "Trick" did sound catchier.) György himself, so notorious for his inability to praise the work of his former students, declared himself surprised to hear the result had been proved in his own lifetime. He died a month later.

"It's very beautiful, no? Always soothing to watch water flow."

Dixon. On the seat next to him. Tolman made a gesture with his head, halfway between a nod and a no. Soothing was the last word that should come to mind. Streamlines and velocities and Navier-Stokes equations —*that's* what one should be seeing in water flow. That's what should be flooding the mind of any young person in the field who wanted to do something, become someone, Tolman felt like admonishing Dixon.

They had run into each other on the railway platform at Hausach. Tolman had been pleased to have been recognized by Dixon, from the University of Arbutus, even though they had never met before. He looked at his fellow passenger now, perched on the edge of his seat like an expectant schoolboy, his feet barely reaching the mat, a thick leather satchel clutched in his lap. Tolman felt an avuncular twinge and remembered his own days as a fresh assistant professor, almost twenty-two years ago.

"Did you know that my advisor was György?" Tolman told Dixon, shifting his large soft body in his seat. "Back when he was still taking Ph.D. students, before he became such a recluse. He used to throw all sorts of impossible problems at us, without mentioning that mathematicians through the ages had tried and failed to solve them." Tolman chuckled. "I remember when we first heard of the problem of flow past a corner. Bramer was a student of György as well, and we both thought it would be an evening's work— we were in our twenties and so naïve then. We ended up racing through grad school trying to see who would crack it first. Look how long it took, though, to finally..."

Tolman stopped and ran a hand over his remaining hair to check that it was in place, the way György used to do with his own. How to convey to this aspiring young colleague all the effort, all the sacrifice, that good mathematics demanded? The exhausting rivalry that had escalated over the years between Bramer and himself over trying to be the first to prove the result? He searched for some sage advice to bestow on Dixon, but was overcome by the poignancy of his own accomplishment.

"There was something I wanted to ask you about your proof." Dixon undid the strap of his satchel and extracted a sheaf of papers.

"Not a mistake, I hope," Tolman said jovially. He decided he liked this Dixon—the quiet personality, the studiousness he projected as he regarded life from behind his thick black-rimmed glasses. Tolman himself had been equally serious at that age—it had been a way of concealing the terrible shyness he felt. He contemplated asking Dixon if he too was shy—would that be too personal a question?

"Actually," Dixon said, and Tolman looked at him, amazed. He couldn't possibly think there was an error.

"Actually, it's Theorem 2.3. I keep thinking there's something wrong with it."

How absurd. The little runt. Didn't he know the exacting review process that *Acta Fluida* employed? Didn't he know the result had been endorsed by György himself? Tolman felt the skin on the back of his neck begin to itch.

"Perhaps we could go over. . . ?" Dixon stopped, a look of alarm beginning to spread over his face.

Tolman forced his jaw to loosen, his cheek muscles to relax. This was exactly the reaction Dr. Winton had warned against, to prevent the eruption on his neck. It had been a perfectly reasonable request. Why had he become so upset? "Roland, you're a mathematician," he heard Dr. Winton say. "It's not as if you're trading companies or landing planes. It's not as if playing with numbers can lead to so much strain." He reminded himself that he was now a senior person in the field. Wasn't it his duty to soothe away the doubts of his junior colleagues? What better way to spend the time in the cab than to give a short, calming lecture on his own work?

"Certainly," Tolman said.

By the time they reached Oberwolfach, though, Tolman was ready to pick Dixon up by the scruff of his neck and give him a good shaking. Could the man really be that obtuse? Part of a mathematician's training was to refuse to take any statement at face value, to be skeptical of every line in a proof. But Dixon seemed incapable of accepting anything as true, asking the same questions repeatedly with such mulish persistence that

Tolman wished he could hasten him along with a stick. Sensing Tolman's displeasure, perhaps, Dixon began to freeze up, and, interpreting this as a further digging in of heels, Tolman felt his agitation turn to cold grinning fury.

"I hope you have a wonderful conference," Tolman said through clenched teeth, and Dixon collected his luggage and fled up the Institute steps.

∞

The Mathematics Institute at Oberwolfach was a low-slung concrete building defined by sharp vertical and horizontal planes, built on a hill about a half mile from town. In its original incarnation, it had been housed in an old hunting lodge, with the purpose of providing technical support for floundering Nazi forces near the end of the war. But for the past several decades, it had enjoyed the reputation of being a leading international center for mathematics. The building overlooked a valley in the Black Forest with a view so breathtaking that it was said that visitors staring at the panorama long enough would feel the solution to any mathematical problem waft in like a breeze off the pine-covered slopes outside.

Tolman drew back the long orange drapes and looked outside his balcony door. The sky was gray and overcast, not a good day to catch breezes. Inside, the room was spotless, but more Spartan than he remembered—a wooden desk and chair in one corner, an unplugged floor lamp, and a low single bed without a headboard, but with a small table attached to it. A mathematician didn't need much, he remembered György saying—a desk, a pen, and a writing pad—the less clutter, the better.

Sometimes Tolman fantasized that he was an ophthalmologist or a surgeon, with sleek, expensive equipment to play with and a progression of glamorous resort conventions to attend. Usually, though, he was happy with the calling he had followed, and today he was not dissatisfied with the room he found himself in. He went over and examined the large metal knob regulating the heating pipe. The gradations clicked smoothly under his fingertips as he rotated it. He turned it this way and that, fascinated with its oiled precision, which to him epitomized all the efficiency and exactness of Germany.

Tolman lay down on the bed to test the mattress and realized how jetlagged he was from the cramped overnight transatlantic trip. He was about to close his eyes, when he remembered Theorem 2.3. He supposed he ought to look at it, to make sure there was nothing to substantiate

Dixon's suspicions. The thought of his own behavior in the taxi filled him with shame. He would search Dixon out at the conference and apologize to him. Perhaps even tonight, though it was probably too soon to sit together, lest Tolman feel like cuffing him again. He turned towards the balcony and watching the sky darken outside, fell asleep.

$$\infty$$

It was the staff that assigned people to the dining room tables, changing the seating pattern at every meal to force the mathematicians to mix. At dinner, Tolman found himself seated at a table with neither Bramer, who hadn't arrived yet, nor Dixon, and mentally calculated the probability that he would miss them on all five days. With ten meals and eight tables, forty-eight attendees and an unbiased staff, it worked out to $(42\times41)/(47\times46)$ raised to the power 10, which he quickly gave up trying to simplify in his head. He was a mathematician, not a calculator.

"Sort of like musical chairs," he said, commenting on the seating to the person next to him, whom Tolman had never met. The man had a large, round head, with an incredibly red face, partly covered by a handkerchief he held over his mouth and nose. He chuckled from behind his handkerchief, so emphatically that Tolman wondered if the reference had been understood.

"I'm Roland Tolman," Tolman said, extending his hand, and the man sneezed twice into his handkerchief.

"Lazarotsky. Like in the book," the man said, using his free hand to shake with Tolman.

"The book?" Tolman repeated, wondering for an instant if he meant the Bible, if there could have been a character named Lazarotsky in it.

"The book on boundary conditions for elliptic operators," Lazarotsky said grimly, his face flushed in insult, "translated from the Russian."

Fortunately, the food arrived just then, wheeled out on carts by the sturdy kitchen women in their blue uniforms.

"Have some soup," Tolman said, ladling out a bowl of thick dumpling soup for Lazarotsky, "it'll be good for your cold." Lazarotsky accepted the bowl with a grunt and did not look again at Tolman.

The main course was endive-wrapped sausage baked in cream and covered with a cheese sauce. Tolman suspected this was not the best antidote to jetlag, but the rumbling emptiness in his stomach made him eat enthusiastically. He even helped himself to one of the three extra sausages that had been apportioned per table. His Finnish colleague Hynoven, whom he had not seen for a while, bought him a beer, and then two more, which helped ease the food down. Tolman felt the combination of beer and cream rise to his head.

Anna, the Canadian woman sitting on the other side of Tolman, was not a mathematician. She was there with her husband whom she introduced as Michel Dresser, an algebraic topologist on a three-month visit unrelated to the conference. Tolman was close enough to smell Anna's perfume even through the meaty vapors rising from the platters on the table. Perhaps it was the angle at which he sat, but everything about her face seemed strangely elongated—her chin, her nose, her lips which seemed to project out to a point as if she was puckering them for a kiss. Even the wisps of her eyebrows arched so high that they seemed ready to lift off from her forehead. She looked like a refined, very elegant collie.

She was also, Tolman decided, the most beautiful woman at the conference. Marie Joly from Belgium could be a contender if she ever bothered about her appearance, the three or four graduate students around were all too young, and the Popescu sisters, who had caused such a stir twelve years ago when they had arrived from Bucharest in their shiny black dresses and luxurious matching furs, seemed to grow tragically more dowdy and bookbound with every conference.

"And what do you work on, Mr. Tolman?" Anna asked with a light French accent, pronouncing the "tol" as "tool."

"My wife's hobby is to collect the names of different fields of mathematics," her husband said.

"It's fluid dynamics. Applied mathematics. Not as abstract as your husband's."

"Which means he's doing things people might actually pay for," her husband interjected.

Tolman blushed. "I'm afraid I only prove things—useless things like the existence of solutions that engineers already take for granted. Besides, it isn't as if algebraic topology is not useful..."

Dresser cut him off. "Really, Dr. Tolman. Shouldn't you leave it to me to take up the defense? As the only pure mathematician around, I insist on this prerogative."

He said it as a joke, but Tolman did not miss the sharpness beneath the surface, like a hidden layer of ice crackling underfoot.

"And how do you spend your days, Mrs. Dresser?" Hynoven asked.

"She's an artist," her husband replied again. "She makes quilts."

"I even have a loom here, believe it or not, to weave cloth," Anna said, apparently unmindful of her husband answering for her. "You should all come and see it some day. It's in this huge open space with a skylight, right outside our room on the top floor."

A vision of Anna alone in a sunlit tower, working on her loom like the heroine of some fairy-tale came to Tolman. He did not know much about quilts (or art in general, for that matter) but felt a need to find out more about Anna. "What kind of quilts?" he asked.

"Little ones, with all sorts of scenes on them. Snow, mountains, trees. I sew them together to make bigger ones. But the biggest I've ever made would still only cover half a single bed."

"*Pure* quilting, in other words, as opposed to *applied*. Runs in the family," her husband said, and the lines of Anna's eyebrows deepened, bringing an irresistible sadness to her face.

Why could she not have been married to him instead? Tolman was sure he could learn to be proud of a wife who made quilts. They would live in a room next to the loom and decorate the walls with her work. He tried to imagine where her husband could have found her, how the two might have met. Except for a brief relationship with a fellow student during graduate school, Tolman had been single all his adult life. The probability that he would ever marry, he knew, was decaying with time. There were so few women in mathematics, and he did not really feel comfortable with people outside his profession. He had once subscribed to a dating service, but had felt ungainly and tongue-tied on the dates they had sent him on. The last woman, a realtor, had pressed him to talk about his work. She had listened attentively as he poured his heart out about existence and uniqueness and turbulence. At the end of the evening she pecked him lightly on the cheek. "In case you're ever ready to buy," she had whispered, squeezing a card into his palm.

The dessert carts were rolled out, and the server, a large woman with red cheeks like those of a doll, placed hearty servings of apple strudel in front of each of them. At first, Tolman decided he would not touch his strudel, but then he thought he would have just a little to appease his stomach, which had begun rumbling again. He ate half of what was on his plate, then half of what was left, and half of what was left again. He remembered his father showing him with a piece of cake. "See, Roland? See how I can take a half again and again, and still never run out of cake?"

He was concentrating on dividing the last few crumbs left on his when he saw Anna with her sad eyebrows staring at him. "Goodnight Dr. Tolman," she said. "I don't know if we will sit together again, but do come up some time and see the quilts."

Later, when some more of the people had left, he caught sight of Dixon and waved to him across the tables in between. By the time Tolman could negotiate his way over to him, Dixon had disappeared from the room. Instead, Tolman joined a table of colleagues gossiping about a husband and wife, both mathematicians, who had worked as a team and who had just divorced. "It was all the papers they wrote together—they never could agree on the notation. Philippa told me how at the end he started using $\mathbb{R}$ to denote the *complex* plane, just to taunt her." Tolman laughed with everyone else and had another beer, so that by the time he got back to his room, the orange curtains were blurring with the white walls. Theorem 2.3 was not even a memory as he fell fully dressed into bed, and into sleep.

$$\infty$$

At 3:45 am, he was wide awake. He forced himself out of bed. He had learnt long ago in graduate school not to let sleep rule him—to regard such early awakenings as gifts, using them to get an edge over other students (particularly Bramer) while they slept. Today, though, he was still a little drunk. In his throat, he felt the bite of acid reflux.

He threw open the curtains and was greeted by the blackness outside. Then the outlines of the mountains emerged, like spirits revealing themselves out of the air. He went and played with the knob controlling the heat, again enjoying the smooth way it rotated, like the dial of a well-lubricated lock. Then he sat down at the desk and opened his paper from *Acta Fluida*. "On the existence of solutions for Navier-Stokes flow in the presence of obstacles with Lipschitz boundaries."

He began scouring it line by line, poking every lemma, prodding every proof, seeing if any step was loose, would give. "Think of a dentist examining your mouth with a pick," he heard György say. "Assume there is a cavity lurking under every innocent surface."

At 5:00 am, he got up to play with the heating knob again. He swallowed some antacid and wished they had coffee in the room, then sat down once more to examine the paper.

Soon after, he started getting a sense of what might be wrong. It was the crucial second step, at the heart of his trick, where he mathematically stretched out the obstacle and its surrounding region to make the corner disappear—the step that always reminded him of drawing a picture on a piece of balloon skin and pulling it into a different shape. He had used a lemma from one of György's first papers, but could that really be applied when the interface was not smooth? Excited by the scent of error, even though it might be his own, he pulled out the György paper and found his attention drawn at once to the continuity assumption. Would it hold in his case? It was a very subtle point, one easy to miss. Could Dixon really have stumbled onto this?

By the time the day emerged from behind the mountains, Tolman had managed to pierce a hole into his theorem and let all the air out of his trick. He felt himself sliding into depression, and clutched at the fading strands of exhilaration he had experienced in the hunt for the mistake. All was not lost, he told himself, he would find a way to repair the proof. Even though Dixon had a hunch there was something amiss, it was unlikely he would be able to pinpoint it. There would be no need to advertise the mistake—Tolman would publish the correction in the guise of a more general proof, and simply stop referring to the original paper. He'd heard that even György had bodies like that buried in his earlier work.

It was almost 8:00 am by now, so he went down to the dining room for breakfast. Even before he entered, the back of his neck started to itch. It was the voice, of course—the abrasive edge, the underlying steel, that the tone, no matter how smooth, could not conceal. There he sat at the far table, surrounded by eggs, bacon and graduate students, and Tolman noted with satisfaction that since their last meeting a year ago, Bramer seemed to have put on at least as much weight and lost as much hair as he had.

Seeing him, Bramer rose and strode over to where he stood, a look of such delight on his face that for an instant, Tolman wondered if he had not misjudged him. Then he remembered to look at the eyes, and saw again the light that danced like mercury in the irises, the pupils that never stood still.

"Welcome," Bramer said, as if he had personally arranged the conference for Tolman, as if Tolman was the guest of honor they had all been waiting for. Wrapping a hand around his shoulder, Bramer led him to the breakfast buffet. "Let's see what we can get you, whether anything looks fresh today." Picking up a pair of tongs, Bramer began rummaging in dissatisfaction through the bacon.

As Tolman ate the toast and boiled egg that had finally met with Bramer's approval, he looked around the room. The tables were all filled now—there was Hyvonen, and Crofton and Eisner—all people he waved at, all people he would have enjoyed breakfasting with. What was he doing stuck in this circle of Bramer's graduate students? Was this a subtle way of belittling him, or was Bramer up to something more sinister—a plot, perhaps, to isolate him? He would not stand for it, Tolman told himself, he would get up and go to another table. But just as he was about to excuse himself, Bramer turned around and smiled at him, and Tolman found himself remaining where he was, and silently chewing the food his colleague had chosen for him.

$$\infty$$

The day passed by in a haze. It was too early in the conference to slip out unnoticed, so Tolman sat in the overheated room, unable to shut out the speakers' voices.

He tried to apologize to Dixon between lectures, but the latter darted away at his every approach with the skittishness of a deer. All day long Bramer insisted on acting the old lost friend, sitting next to him wherever he went, and Tolman had to be careful not to write anything about Theorem 2.3 on his notepad, lest it be noticed. Most exhausting were the comments on the lectures that Bramer kept whispering like a gossipy schoolgirl into Tolman's ear. "He's been giving the same talk for two years." "What a ridiculously broad assumption to make—why not just assume the entire theorem?" "Cunha proved this years ago—looks like the French are just rediscovering it." Little nips and jabs at every speaker, offered as implicit evidence of trust and friendship, which only served to remind Tolman of the unease he had felt around Bramer ever since graduate school.

That evening, when Tolman found himself seated right next to Bramer at dinner (the probability being 1 out of 23.5), he realized how easy it must have been to slip in beforehand and move place cards around. Bramer started waxing nostalgic about their student days, dredging up anecdotes about György and Galena and all the other professors. "Remember Bartle?" Bramer said. "Remember how we all skipped his Thursday class and went to the Craig Street Inn to have a beer?"

"And the next week, Bartle himself came with us," Tolman said, joining in despite himself.

By the time dessert rolled around, Tolman had consumed the same number of beers as the night before, and realized that he had been much too harsh on Bramer, who was, after all, his friend. He looked around for Anna, wanting to point her out to Bramer, but didn't see her. He had tried going upstairs to the top floor after the day's talks, but had not found the lobby with the skylight she had spoken of. "She's really beautiful," Tolman said. "Though she looks so sad. Who knows what she sees in her husband?"

"I met someone interesting at the conference as well," Bramer said, pouring more coffee into their cups. "Just before dinner. From the University of Arbutus—his name is Dickens or something—do you know him?"

Tolman felt his head suddenly clear. It couldn't be, he thought. Even Bramer couldn't work that fast.

"Dixon. That's what it is. Dixon. He says he shared a cab with you on the way here."

So this was it. The friendliness and goodwill that Bramer had been lavishing on him all day—it had been a ruse to get his guard down. Tolman braced himself for the kill.

"He seemed intelligent enough, until he suddenly started making the most preposterous claim. About the obstacle paper you wrote. Such a beautiful, beautiful paper, I must say—people ask me to this day how far I came along in proving the result myself—and of course I tell them it doesn't matter any more, Tolman beat me to it. But this Dixon character—for some reason, he's convinced himself he's found a mistake—Theorem 2.3 I think he claims."

"Yes, he made some such remark to me as well." Tolman tried not to let his voice waver. "In the cab. He was hardly coherent."

"Completely absurd—and I told him that, of course. In fact, I even offered to go over the proof with him—I'm meeting him tomorrow. Who even knows where this obscure University of Arbutus is—never heard of it. All these people coming to conferences these days—I wonder what idiots in university administration are funding their trips."

Bramer began to rant against his own higher-ups, and Tolman tuned him out. There was no time to lose, he realized, bringing the cup to his lips and sipping the coffee Bramer had poured. He had to get to his room at once and fix the proof before his old rival did.

∞

By Wednesday, Tolman was in a panic. Nothing had worked. Pages of useless equations lay strewn over the room, like the droppings of some unroped mathematical truth. His eyes hurt and his body ached from sitting at the desk in his room all day Tuesday. He had gone downstairs only for meals, and seen Bramer and Dixon busy scribbling on paper napkins each time. All night, he had lain awake in bed, churning mathematical arguments in his head, until at 4:00 a.m, an exhausted sleep had overtaken him. This morning, he had awakened with a sore throat and runny nose. Viruses took a week to incubate, he had reassured himself— it couldn't be Lazarotsky passing on his cold in revenge.

He almost didn't go for the traditional Wednesday afternoon hike, but then decided the walk would clear his head. The last time he had attended a conference at Oberwolfach, their mathematician guide had dragged them over mountains and through valleys in search of authentic Black Forest cake, a search that had turned into a seven-hour ordeal when they had not been able to find their way back. This time the organizer had less ambitious plans—he simply led the way along the river towards town, where there was a small fair going on. The mathematicians followed along obediently, several of them in sturdy hiking boots packed specially for this afternoon.

Tolman walked silently near the back of the group, staring at the clean, cold water. They were coming up to the stub of snow he had seen from the cab; their path would take them only a few yards from it. He let the dawdling Popescu sisters, their furs bedraggled by the years, pass him. Then he walked over to the river bank.

It was, indeed, a bridge that had once stood there. All that survived now was a stone abutment built into the bank and a short piece of walkway that ended abruptly where the rest had fallen off. The snow crunched and packed under the smooth soles of his shoes as Tolman edged his way along the walkway and peered over the side. Below, the flow was an example of almost textbook perfection, one that he had drawn so often in class on the blackboard. Water swirled around the right-angled corner of the support, the froth curling into tiny eddies as it raced to join the river beyond.

Tolman got down on his knees, then stretched his body flat against the snow. He pulled his head up to the edge and gazed at the water.

When he was ten, his family had moved to a house next to a wooded area through which a stream ran. He would lie alone on the bank for hours on end and watch the twigs and leaves float by. Patterns would flower and fade like in a kaleidoscope, as they were doing now on the surface below.

How many equations, he wondered, were being solved at each instant to form these patterns? How many variables, how many parameters, how many pieces of data were involved? How many decades before mathematicians had the proficiency to prove theorems assimilating them all? The models he looked at were derived after many assumptions and all sorts of simplifications. Even those models were so inscrutable, so resistant to the pencil and paper manipulations of his tribe.

And who cared, anyway? What was the point of being a mathematician? Of spending years of his life trying to validate solutions that scientists took for granted, that engineers routinely approximated? The real world needed ships and dams and planes and rockets built. It demanded computer-generated approximations in seconds, color maps in two and three dimensions. It did not time its business to the musings of mathematicians any more.

Tolman felt the snow melt against the warmth of his legs and soak through the cloth. He wished György were here. György rallying him to get up and resume the fight. György reminding him of rockets that had crashed, oil platforms that had collapsed. Of all the catastrophes that could have been averted precisely if the world had paid attention to its mathematicians. György telling him how he was indispensable, how without mathematicians there would be no science or engineering or civilization, how Nature would reign, cruel and untamed and uncontested.

He was trying to remember György's green-blue eyes, his carefully combed strands of white hair, his Hungarian accent, when he heard footsteps and turned around. It was Bramer, stepping cautiously over the stones towards him.

"Oh, there you are," Bramer said, as if, while out on a casual stroll, he had come quite naturally across Tolman sprawled on the snow.

"I was trying to study the water under the bridge." Tolman made an awkward, unsuccessful attempt to get up. "It's hard to look at a river and not think of György."

Bramer offered a hand and helped pull Tolman to his feet.

"He'd have loved this spot, wouldn't he?" Bramer brushed some snow off Tolman's coat. "I remember the time he took me down to the Monongahela. We spent the whole afternoon looking at streamlines. His wife packed egg sandwiches for him, which he hated, and gave to me. I can't believe it's been two years. I still dial his number when I'm stuck on a proof, not remembering he's gone."

They stared silently at the bridge, at the water.

"A wonderful place for the obstacle problem, is it not?" Bramer said. "One could even use the trick."

Tolman noticed it at once, as he knew Bramer must have meant him to—the omission of his name in prefacing 'trick.' He shivered in his damp clothes and waited for what Bramer had come to say.

"Look at you, you're all wet. We better head back—I'd hate to see you catch a cold. We can have some coffee, talk a little. Maybe even discuss your obstacle paper."

Suddenly Tolman felt a violent need to sneeze. He pulled out his handkerchief from his front trouser pocket and buried his nose in its folds, but the dampness of the cloth only exacerbated his urge.

"It's the continuity assumption, isn't it?" Bramer was saying. "Dixon pinpointed it for me. It's such a useful theorem, too – it would be a shame to lose it. Perhaps we could all sit down together and look at our options."

Our options. The pronoun arrested the sneeze that was preparing to discharge through Tolman's nostrils, stopping him in mid-inhalation.

"He's smart, this Dixon guy. He feels we might be able to save it."

It was difficult to assimilate the full arrogance of Bramer's proposal quite at once. All Tolman's insight, all his years of research, and now Dixon and Bramer were to be his saviors? So that they could rename his result the Bramer-Dixon-Tolman Trick?

Then a more frightening thought occurred to him. What if he didn't join in, and they managed to correct the theorem before he did? He could hardly forbid them from working on it. What if it became known as the Bramer-Dixon Trick?

"I was thinking we could meet Dixon for breakfast tomorrow."

For an instant, Tolman imagined grasping Bramer by the arm and swinging him into the river from where he stood. Watching his colleague float away through the ice chunks, flipping from side to side as he hit the banks. What would be the appropriate boundary data when Bramer made contact with an edge?

Then Tolman felt ashamed. Bramer was who he was, and there was technically nothing dishonorable in what he had said. An uninformed observer might even characterize his overture as benign, rather than the raw attempt to hijack the theorem that Tolman recognized it as. There was nothing to do but politely decline the invitation, and hope that Dixon didn't by some crazy fluke stumble onto a corrected proof before he did. Tolman had, after all, been the person to come up with the original idea, so there was reason to believe he would be the one to correct it as well. This was not the moment to lose faith in himself.

"Perhaps some other time," Tolman started to say, but was overcome by a volley of sneezes.

$$\infty$$

By the time they got back, Tolman was feeling quite unwell. His head hurt, his nose ran, and the rash on his neck seemed to have moved to the inside of his throat. He thought he had a fever, but didn't know whom to contact at the Institute for a thermometer. At dinner, he forced himself to eat some of the pea and sausage soup, but the main course of stuffed cabbage sent him scuttling back to his room. For a moment, he actually contemplated making a stab at Theorem 2.3. But then he turned the knob on the heating pipe up and down a few times, adjusting the temperature, and with the soup roiling in his stomach, climbed into bed.

Around midnight, he checked his pulse. It was racing along at around a hundred and twenty. He tried gauging the temperature of his cheeks and forehead with the back of his hand, but everything felt equally hot. He got up and turned the heat down a notch. An hour later, he began shivering and had to turn the heat back up. Still later, he began to feel once more as if he was boiling and threw off all his covers. In his delirium he imagined that he was back at the Wolf River. In his hand was a shovel, and he was going to dig up the remains of the bridge to smooth out the flow. But there was no way to climb down to the level of the water. So he took off his shirt, then the rest of his clothes, and jumped in. The water felt cool and refreshing against his burning forehead, his sweltering skin.

There was a knocking on his door the next morning. Tolman ignored it. Hyvonen looked in at lunchtime and came back with some rolls and butter for him. He stayed in bed all day, popping aspirin whenever his pulse rose above 120. Once, when he began thinking he was at the Wolf River again, he soaked some water on a handkerchief and laid it across his forehead.

Around four, he fell into a troubled sleep for an hour and dreamt of György. When he awoke, he thought György was disclosing the corrected obstacle proof to Bramer in a corner of the room. He called out to the shadows in protest, but there was no reply.

At six-thirty, he decided he had to eat a proper meal to keep up his strength. He forced some clothes over his aching body and went down to the dining room. Dinner was a blur. There was a potato and pork stew, flavored with curry powder for a reason he was unable to understand. He was startled to find a section of starfruit nestling radiantly between the slices of pork on his plate. He wondered if he was delirious again, and bit into the star-shaped piece. It squirted curry and pork fat in his mouth.

He was about to drag himself back upstairs when Holzerman, one of the organizers, turned to him. "How is your theorem doing?" he asked, and the voices at the table died down. "Bramer said there might be a mistake somewhere."

Tolman felt the room lurch. "I'm working on it," he said. "Trying to," he added weakly.

"Your talk's tomorrow morning, isn't it?"

"I hope I'll be better by then." He waited for the conversation to pick up again, then fled.

Bramer came to his room at 9:00 pm, just as the fever, fueled perhaps by the curry or the starfruit, seemed to be surging back.

"Don't worry about tomorrow, don't worry about anything" he said. "I asked Holzerman to let Dixon give a talk in your stead. He agreed, so now I want you to take complete rest."

Tolman began mumbling in protest, but Bramer shushed him. "We're still working on the proof. If nothing else, we'll at least be able to explain where the argument in your paper breaks down.

"Now remember. Rest," Bramer said, placing a hand on his forehead, and Tolman tried flailing out but his arms were too heavy to lift off the bed.

$$\infty$$

The night was worse than the previous one. Once again Tolman found himself in the river, hurtling towards the bridge this time, surrounded by millions of fluid particles. The stone came up with tremendous speed, and his body bounced off the right-angled edge. He whirled through the water, spinning like the ball in a pinball machine, then was thrown into an eddy and went spiraling to the river bottom. Perhaps he drowned, because he suddenly experienced a remarkable buoyancy, and saw himself afloat on the river surface, his spread-eagled body being borne downstream.

When he awoke, sunlight was soaking through the balcony drapes, bathing the walls in an orange sheen. He looked at the bedside clock—it was a little past 10:30 am. Dixon, he realized, must be in the middle of his lecture right now. He wondered if he should throw off the covers and stagger through the snow to the conference room, burst into Dixon's talk and take the podium from him. What would he say, though? The transparencies he had brought along were teeming with references to the Tolman Trick. He imagined Bramer at the back of the room, standing up to repudiate him. The tale being told for years to come about Tolman going down in flames at Oberwolfach.

No, emotion was always the enemy of good mathematics, as György used to say. He would stop thinking of the problem for a while, let his mind float free to allow new ideas to drift in, come back and fight another day. It was his fear and nothing else that had so magnified the specter of Dixon having already corrected the proof. He had to keep reminding himself that it was much easier to find a mistake than to repair it.

Tolman sat up in bed. He still had a fever, he could tell, but his pulse was slower. He felt hungry so he put on his clothes. As he stepped outside the door he remembered that breakfast had ended more than an hour ago. A woman in a blue uniform looked up from besides a bucket on the floor. "Guten Tag," she wished Tolman, then went back to scrubbing the tiles vigorously with a brush.

Tolman made his way around the giant chessboard on the hallway floor. He walked past the bookshelves built into the walls, lined with the great classics on mathematics. Newton and Hilbert, Euclid and Descartes nestled reassuringly next to one another.

He came to a narrow wooden door between two of the bookshelves. He had not noticed it before and wondered if it was an exit reserved for the staff. The blue-uniformed woman was oblivious to him, her body rocking back and forth as she attacked the floor with her brush. Tolman opened the door and squeezed through.

He stepped into a corridor and followed it to a vestibule from which a spiral of steps rose. Light shone down from the floor above. He mounted the stairs and stepped onto the polished floorboards of a large hallway. Sunshine poured in from the windows on either end, and blazed through a skylight on the far side.

Beneath the skylight what appeared to be clothes had been spread out on the floor, perhaps to dry in the sun. Tolman moved closer and saw that they were in fact quilts, some no larger than a table napkin, others the size of a child's cot. Scissors and bits of cloth lay strewn about, and a measuring tape basked uncoiled along a window sill. On a table by the wall was a loom, with threads of different colors plunging to spools on the floor.

Tolman bent down and examined the quilt closest to him. It was made up of individual patches, each about a foot square, arranged in three rows of two. Each patch had a miniature scene depicted on it—mountains of white snow, forests of pure green, clouds floating against backdrops of blue sky. He moved from quilt to quilt, and saw roads and rivers and villages and lakes, even cows grazing in a field. He picked up a patch from the table and felt it between his fingers—the cloth had a cottony feel to it. Sewn into the material was a blue ribbon of river, running unimpeded across a snow-white landscape.

Tolman ran his thumb across the curve of the river bank. He could not detect any kinks, the boundary between blue and white was smooth, graceful. He was examining the stitching on the back when he heard the turn of a doorknob. Anna appeared in the hallway and gasped when she saw him.

"Oh, Dr. Tolman. You startled me. I didn't know anyone was here." She stopped, noticing his feverish eyes, his unshaven face. "Are you all right?"

"Yes, don't worry, I'm okay. I'm really sorry. I didn't mean to intrude like this. I just saw the stairs and..."

"No, no, that's fine. I told you to come, remember? I'm glad you found it."

Tolman stood by the table and smiled awkwardly. Anna's face was long and expectant, waiting for him to speak. "It's very beautiful," he finally said.

"All of it. Your. . . art." The word felt apt and full of possibility in his mouth. He had never made such a judgment, uttered such a pronouncement, paid such a compliment before.

"Thank you. Dr. Tolman." Anna lowered her eyes to the square in his hand.

"I'm especially fascinated by this one," he said, holding it up. "How serenely the river seems to run." Tolman touched a fingertip to the blue. "Tell me, were you on the hike into town?"

Anna shook her head.

"There's this point on the river where there used to be a bridge of some sort. Where the water makes its way around an old stone abutment."

Tolman stopped. Anna was looking at him, her silence encouraging him to go on. Sunlight dappled her hair and slid down the slopes of her face. Her mouth extended towards him, her lipstick darkening its edges. The lines of her eyebrows reached high towards each other in anticipation. She was waiting, he knew, for an explanation, an elaboration.

How could he begin to comment on the chasm between her river and his own? To express to her the naïveté, the brilliance, the startling beauty of what she had sewn? "Where the water..." he began, then broke off again. The innocence in her eyes was something he couldn't bear to remove. Behind it was a clarity he realized he would never possess.

"It's beautiful, isn't it? The river," Anna said.

Breaking Down the Barriers

"Hey, Ben, come here. I need your help in the kitchen."
"What are we going to do, Daddy?"
"I need to practice for Mrs. Anderson's class tomorrow."
"OK."
"Wash your hands first."

Ben scampers to the bathroom, holds his hands under the running water for two seconds, and hurries back. The youngest of my four sons, Ben has just turned seven. He joins me at the butcher-block table, climbing up on a wooden stool so he can see more easily. He's inherited my father's large head, and his wispy blond hair flies in all directions. His blue eyes shimmer with anticipation. He's a cute kid, I think. He's also a very wet kid.

"Ben, please go back to the bathroom and turn off the faucet. And don't forget to dry your hands."

"But there's no towel in the bathroom."

He's right. I forgot, once again, to put a hand towel in the bathroom. I grew up thinking that hand towels, usually decorated with flowers or butterflies, were the responsibility of the woman of the house, but I've discovered that even this all-male, post-divorce household needs hand towels.

After I give Ben a towel (it's actually a dish towel, but we don't care), he returns to the kitchen table. He's my test subject, helping me practice a chemistry experiment I plan to do in his first-grade classroom tomorrow.

I pour enough whole milk in a cereal bowl to cover the bottom of the bowl. Then I get out small vials of food coloring.

"Daddy, let me do it."

"I'll do the first one. Then you can try. You have to squeeze the little bottle very gently. See. Like this. You want to let just one drop of coloring fall into the bowl... OK. I put a green drop in one corner of the bowl. Do you want to try it with the blue?"

"I want to do the red. It looks like blood."

"OK. Be very careful now. Just give it a little squeeze. Let one drop fall into another corner of the bowl."

I reach out to help Ben with this delicate operation, but he pushes me away.

"I can do it. Let me do it."

Ben squeezes the bottle, and five drops flood into the bowl. Two more land on the table.

"Oops."

"Ben, let's wipe up the mess. Then we'll try again."

I wipe the food coloring off the table, but the red stain has already started to seep into the wood. Oh well, it's a small price to pay in the service of science.

"Hey, Daddy. Look at me. I'm bleeding!" Ben laughs, showing off his red-stained fingers.

Maybe it wasn't such a good idea to let Ben handle the food coloring. He hasn't developed those fine motor skills yet. But that's just the sort of information I was hoping to gain from this practice session. It's one thing to let my own kids walk around stained, but I'm not sure if Mrs. Anderson will be pleased with a whole class of red hands.

I take firm control of the food-coloring bottles, and we start over again. Ben pours the milk this time, not spilling a single drop. I carefully place four drops, each of a different color—red, yellow, green, and blue—into the four quadrants of the bowl.

We stare at the drops. They're just sitting there, suspended in the milk. Each drop is an island.

"Cool," says Ben.

The first time I did this experiment for myself, nearly ten years ago, I was surprised that the drops of food coloring didn't mix with the milk. I had expected a liquid kaleidoscope of color or a bowl of boring brown.

I hadn't considered, however, the fact that milk and food coloring are different liquids and that there might be some barriers to mixing them together. I knew the old adage that oil and water don't mix, but I hadn't considered that lots of other things don't mix either. Milk, with its fats and proteins, behaves a lot like oil. And food coloring is more than 99% water.

Gazing now at the four isolated drops of color in the bowl of milk, I realize that my whole life is full of things that don't mix easily with each other. Why am I so surprised? If oil and water don't mix, why should I have expected work demands and family responsibilities to mix easily? Or chemistry and creative writing? Or the daily concerns of an at-home Dad and a working Mom? Or my dreams of Nobel Prizes and my dreams of family picnics?

As Ben and I stare at the islands of bright color floating in a sea of white milk, I can feel the isolation of the individual colors.

"Daddy, is that it? Are we done? Can I go play with my Legos now?" Ben has apparently not wandered off into a metaphorical wilderness like his father. He's spent ten seconds gazing at a bowl of milk. That's more than enough for him.

"No, Ben. The best part is coming. Can you grab that dishwashing liquid by the sink?"

Ben grabs the bottle of Joy. "Hey, look at me. I've got Joy. She's yellow and smells like lemons."

Joy is the name of the woman who served as a nanny for my kids since we first moved to Minnesota from Washington, D.C., eight years ago. While Ben's Mom was working long hours and traveling extensively as a consultant and I was trying to carve out a part-time career as a freelance science writer, Joy would provide childcare during the day. Now that the kids are all in school during the day, Joy still comes several evenings a week to help me cope with the chaos of sports practices, dinner, homework, music lessons, and community meetings. In honor of my kids' affection for her and her indispensable role in the household, the dishwashing liquid of choice in our family has always been lemon-fresh Joy.

"Squeeze out a drop and let it fall right into the middle of the bowl," I tell Ben. "Be careful. Just add one drop. See what happens."

Before Ben can try it, we're joined by two creatures attracted by the sense that something important is about to happen.

Midnight, our black cat, jumps silently up on the kitchen table. He eyes the bowl of milk, apparently ignoring my iron-clad safety rule that "we don't eat or drink chemistry experiments."

Nelson, my nine-year-old, brandishes a plastic light saber. "I am Darth

Vader. Surrender to the Dark Side," he growls in his deepest stage voice.

"Nelson, look what I did," says Ben.

Nelson and Midnight peer into the bowl.

"Cool," says Nelson, still using his Darth Vader voice.

Midnight keeps his thoughts to himself, but I know what he's thinking. I quickly grab him so can't ruin our experiment or turn his pink tongue to a scary shade of bright blue. I put him on the steps heading down to the basement and close the door.

Ben explains our experiment to Nelson, with confident authority. Nelson's third-grade class will do the same experiment next week.

"OK, guys, it's time to add the soap," I proclaim. "But you've got to promise not to tell the kids in your class what happens. I want them to see it for themselves."

Ben squeezes out a drop of soap. It plops into the middle of the bowl. Immediately, the colored drops recoil from the center of the bowl, fleeing to the sides of the bowl, trailing smears of color. The islands spread and lose their sharp definition.

"Awesome!" Ben yells. As we watch, the colors swirl ever so slightly, approaching each other, yet remaining separate.

"Can I try it?" Nelson asks, forgetting for now that he's Darth Vader.

We set up another bowl of milk, and Nelson gives it a try. Planning to improve on his brother's performance, Nelson adds the drops of red and blue together, hoping for a purple drop. Then he adds two yellow drops on the other side of the bowl. "Look," he proclaims. "Minnesota Viking colors."

"Look at mine," Ben begs. "It's alive."

The colors in Ben's bowl are still undulating and swirling, even several minutes after he added his one drop of Joy.

"Let's try it again," I suggest. Opening up the refrigerator, I pull out a half-gallon container of skim milk and a pint of cream. "We're going to see if we get the same results with these."

Before long, the kitchen table is covered with bowls—eight of them, each filled with a rainbow of colors.

"Whoa! What's going on here?" William, my sophisticated seventh-grade son, enters the kitchen. He's seen quite a few of my chemistry activities before, but I don't think he's seen this one. He's very curious, but needs to maintain his nonchalance.

"That's a lot of milk." Pause. "That's a lot of food coloring." Pause. "Let me guess. This is a chemistry experiment." Long pause. "OK, I give up. What are you doing?"

Now we've got William hooked, too. Ben, Nelson, and I set up the demonstration for William. Instead of adding a whole drop of Joy this time, however, I take a toothpick and dip it into a drop of soap. Now, I've got a thin layer of soap coating the tip of the toothpick. I touch the toothpick to the center of the bowl of milk, wait for the swirl of colors to engage my seventh-grader's jaded mind, and prepare to launch into an explanation.

∞

I plan to give my boys a simple, straightforward explanation of how soap works. But as I begin to describe the chemistry of milk and soap, my mind's eye travels into the bowl of milk. The rest of the world disappears, and I'm swimming in a sea of molecules, a snorkeler suspended above a coral reef. Clusters of whitish fat molecules bob on the waves, gently parting as I glide forward. A team of water-soluble vitamins darts around me like a school of angelfish. Beneath me I spot a large protein molecule; it's a majestic stingray winging its way home.

While reading about and practicing this chemistry experiment the past several days, the beauty and complexity of milk and soap have awakened the alchemist, poet, and philosopher in me. The splendor and order of the physical world hint at lessons that might apply on a more spiritual plane, if only I can distill their meaning. Yet even as I feel the connection between the physical and spiritual worlds strengthening in my mind, I'm aware of another presence. A stern chemistry professor stands resolutely in the rational corner of my mind, arms tightly folded on his white lab coat. "Just stick to the chemistry, Randy. You're a scientist."

Milk is actually a very complex food composed of over 100,000 different molecular species. Seen through a microscope, it's made up of little round spheres floating in a sea of water. The little round spheres are globules containing fat (and also oils, proteins, sugars, and fat-soluble vitamins). These fat globules are hydrophobic ("water-fearing") and don't like to mix with water. Because the globules are less dense than water, many of these spheres rise to the surface of the bowl of milk.

The dyes in food coloring are hydrophilic ("water-loving"), so they don't dissolve into or combine with the fat globules. When the drops of food coloring are added, they find themselves trapped in the crowd of fat globules on the surface of the milk.

Soap molecules have a unique property. One end of the molecule is hydrophobic, and one end is hydrophilic. Like a seasoned diplomat at a summit meeting, a genial hostess at a cocktail party, or a parent mediating a sibling quarrel, the soap molecule can operate in both worlds at once, bringing together molecules that would normally avoid each other. When we dip the soapy toothpick in the bowl, the soap molecules quickly spread across the surface, changing the structure of the milk and altering the surface tension. In the presence of the soap molecules, many of the fat globules collapse and mix freely with the water, thus freeing the food coloring to also mix freely.

That's how soap works. It can break down a stain of fats, oils, or proteins and allow these materials to dissolve in water. It can break down individual barriers and encourage cooperation. It can bring bitter enemies to the bargaining table and restore peace throughout the land.

In the bowl of milk, the isolated colors now dance, play, and merge. Their yearning for community has been answered. They join together, ready to paint a picture of beauty and truth.

All these miracles result from the simple fact that soap is able to operate at the same time in two worlds.

As I launch into this archetypal tale of enemies, battles, and heroic deeds at the molecular level, I can see my children's eyes begin to glaze over. Ben slips off his stool and sneaks away to his Legos. Nelson picks up his light saber and heads off to slay a Jedi Knight. Only William remains; he wants to try a few experiments on his own now that he's endured Dad's lecture.

Erik, my teenager, pokes his head up from downstairs and stares suspiciously at the bowl lineup. "When's dessert? Did you guys already eat the ice cream?"

All the bowls that we own are spread out on the kitchen table, full of different combinations of milk, food coloring, and soap. The colors in the bowls look relaxed and at peace. They've been mingling with each other, finding some balance and harmony. I wish my life felt like that.

I don't want to disturb these works of art, but reality is intruding. I have to supervise homework, finish the laundry, and revise an article that my editor expects by noon. It's going to be another exhausting evening.

I had foolishly hoped to find a quiet moment to read a chapter in that self-help book recommended by someone in my divorce support group. I'm sure the chapter would have discussed the importance of making time for myself. Maybe tomorrow...

My kids are going to be clamoring for dessert very soon. If I put the bowls in the dishwasher, the wash cycle will take 30 minutes, and the bowls will then be too hot for ice cream. I'm going to have to wash them by hand. I'm glad I bought an extra bottle of Joy earlier today.

As I ease my hands into the warm, soapy water in the sink, I think about my Dad. His self-appointed job after every family dinner was to wash the dishes. Every evening, he'd be standing at the sink, engaged in his ritual. First, he'd put on an apron and pull on yellow rubber kitchen gloves. Then he'd scrape the plates, rinse them off, and stack them on the counter to the right of the sink. After running the garbage disposal, he'd plug the sink with the stopper and start filling it with hot water. He'd squirt in the dishwashing soap—Ivory liquid from an opaque white bottle. The glasses and silverware would be washed first, then the plates. Finally, he'd tackle the pots and pans. As each item was washed, he'd dunk it twice in a sink full of hot rinse water and then stack it carefully into the rack to the left of the sink.

He never left the dishes to air dry. Sometimes he'd turn on the radio and dry the dishes himself, excusing me to do homework or play ping-pong with my brother. Sometimes he'd ask me to dry them as he washed. As we worked side by side, he'd talk to me about school or tell me a story from his childhood. I secretly enjoyed these chances to talk with him, but I never let it show.

This short period after dinner was one of the best times of the day. Order was being restored to the kitchen and to the world. My father was performing one of his well-defined roles. During the day, he worked as a university professor. In the evening, he washed dishes in the kitchen. On weekends, he mowed the lawn. In the summer, we took a two-week camping trip, and he washed the dishes in two green plastic tubs. The water for dishwashing (and bathing) was heated over the Coleman camp stove, and he still used the Ivory soap from the white bottle. His world—and my world—was solid, ordered, and predictable.

What happened to that world?

My eyes blur just a bit. It must be a chemical reaction to the soap. I reach for another dirty bowl and continue washing.

Paul Zimet

Star Messengers

Prologue

An isolated spotlight comes up on the aged and blind Galileo Galilei. His hand rests on a shoulder of the Accordion Player who accompanies his song.

Left: Will Badget as Galileo; right: Christopher Hayes, accordionist. *Star Messengers*, Smith College, April 2001. Photograph by Jon Crispin.

Galileo

When I showed the philosophers my glass
and taught them how to point it at the night

they saw nothing—only black and specks of white—
till I translated what they saw:

"That point of brightness... that is Jupiter...
those tiny smudges hugging it are its moons"

Some, not wanting to seem stupid,
would say "Ah yes, I think I see."

While others screwed their eyes and slyly asked,

"How can we know those fuzzy dots are not
objects placed by you inside the tube."

Back then, it angered me.
Now I wonder, were they right?

For me, all phenomena are black
and still I see those points of light
Are they painted in my skull?

My world has shrunk to this rough table
the coarse crust in my mouth
my stomach's growl
my breath
And still the firmament flares bright.

In my unending night the moon mutates
from a pale sliver to dazzling sphere,
the Medicean stars hide and reappear,
Venus veils herself with shadows
then unclothes her limpid beauty...
All instantly.

And even more...
things I'd never seen before
flash inside this darkened chamber

traces, patterns with no key
and like those old philosophers
I say "Ah, yes I think I see."

Spotlight comes up on Johannes Kepler. The accompanying Cellist stands near him.

Left: Stephen Katz, cellist; right: David Greenspan as Kepler. *Star Messengers*, Smith College, April 2001. Photograph by Jon Crispin.

Kepler

The planets and the stars
all blur.
I'm a dog without a nose,
a cook who
can't tell salt from sugar.
How ridiculous to
choose to be
a nearsighted
astronomer.

I had no choice.
I need proofs
that the patterns
I divined
existed not just
in my mind—
that what I saw
with my closed eyes
illumines Heaven's mysteries.

Kepler raises a telescope to his eyes.

Biographical Notes

Marco Abate is a Professor of Geometry at the University of Pisa, with research interests ranging from holomorphic dynamical systems to complex differential geometry to geometric function theory, and writing textbooks as well. That's by day. Nights, he writes comic book stories which have been published by major publishers in Italy (and recently also prose stories). If you feel an irresistible urge to find out more about him, he invites you to—but does not vouch for—http://www.dm.unipi.it/~abate.

Colin Adams is the Thomas T. Read Professor of Mathematics at Williams College. In addition to research papers on knot theory, he has written *Why Knot?*—a mathematical comic book with attached toy. Among his honors for teaching and exposition, he was the Pólya Lecturer of the Mathematical Association of America for 1998–2000. The MAA also published his humorous DVD (joint with Thomas Garrity) *The Great Pi/e Debate*. He writes a regular humor column for *The Mathematical Intelligencer*.

Madhur Anand holds the Canada Research Chair in Global Ecological Change at the University of Guelph—an exceptional opportunity to carry on her studies of forest ecology. Her poetry has recently appeared in literary journals including *CV2*, *The New Quarterly*, and *Room*. She is currently working with Adam Dickinson on editing an anthology of poetry about restoration ecology.

Sandy Marie Bonny has an undergraduate degree in geology and graduate degrees in earth and atmospheric sciences. She is now teaching part-time at the University of British Columbia while she puts the finishing touches on her first collection of short fiction. Her stories have been published in various literary journals, including *Grain, en Route,* and *Spring,* and have aired on CBC Radio. Her story *mandala* placed second in the short fiction category of the 2001 Canadian Literary Awards.

Wendy Brandts is a scientist (research in theoretical physics and biology at Universities of Toronto, Oxford, and Ottawa) and writer. Her fiction and poetry have appeared in *Descant, This Magazine, Room, Collected Iron Works, Wascana Review, Room of One's Own, Carleton Arts Review,* and elsewhere, and have won several awards.

S. Isabel Burgess is currently a PhD student in high-energy physics at the University of Toronto. She also holds a B.A. in creative writing and sociology, and an interdisciplinary M.A., both from the University of Victoria. Her work has previously appeared in *The Claremont Review, Grain Magazine, PRISM International, The Malahat Review, Prairie Schooner,* and *The New Quarterly*.

Robin Chapman is Professor Emerita at the University of Wisconsin Madison; her scientific work is in language development in children with Down syndrome. Her poetry has appeared in five books, and in many magazines including *The American Scholar, The Hudson Review, Fiddlehead, Poetry,* and *Southern Review*. Her book *Images of a Complex World: The Art and Poetry of Chaos* (2005) intersperses her poems with fractal art of J.C. Sprott. Fittingly, she co-edited the anthology *On Retirement: 75 Poems* (2007).

Chandler Davis has been on the faculty of the mathematics department at the University of Toronto since 1962. He has been an editor of *Mathematical Reviews* and a Vice-President of the American Mathematical Society. His mathematical research ranges quite widely. His (non-mathematical) prose has appeared in *Astounding Science-Fiction, The Nation,* and elsewhere; selecta will be published as *It Walks in Beauty* (edited by J. Lukin; Aqueduct Press). His poetry has appeared in *Canadian Forum, Saturday Night, This Magazine,* and various little magazines. Combining his personae as mathematician and wordsmith, he has been on the editorial team of *The Mathematical Intelligencer* since 1987.

Florin Diacu is Professor of Mathematics at the University of Victoria, and editor of three mathematics journals. His *Celestial Encounters: The Origins of Chaos and Stability* (with Philip Holmes) won the Best Academic Book Award in 1997; his *The Lost Millennium: History's Timetables under Siege* is a best-seller; his next book, *Megadisasters: The Science of Predicting Natural Catastrophes*, is coming out in 2008. He is a voracious reader and is fluent in five languages.

Adam Dickinson's poems, articles, and reviews have appeared in a number of journals in Canada, the UK, and the USA. He is now Assistant Professor of poetics at Brock University in Ontario, where he also serves as co-editor of the literary journal PRECIPICe. His poems have appeared in two acclaimed books, *Cartography and Walking* (2002) and *Kingdom, Phylum* (2007), and he has been anthologized in *Breathing Fire 2: Canada's New Poets*, *Post Prairie*, and *The Echoing Years: An Anthology of Poetry from Canada and Ireland*.

Susan Elmslie's first collection of poetry, *I, Nadja, and Other Poems* (Brick, 2006) won the A.M. Klein Poetry Prize and was shortlisted for other prizes. Her poems have also appeared in several journals and anthologies, and in a prize-winning chapbook, *When Your Body Takes to Trembling* (Cranberry Tree, 1996). She has been a poetry Fellow at Hawthornden Castle in Scotland. She currently teaches at Dawson College in Montreal.

Claire Pratt Ferguson is an artist and writer; she frequently writes about the art of her husband, the mathematician and sculptor Helaman Ferguson. She has written and lectured widely on his art and the mathematical ideas that inspire it, including the prize-winning *Mathematics in Stone and Bronze*; and she has often curated his exhibits, including several at major meetings of mathematical societies.

Emily Grosholz is Professor of Philosophy at the Pennsylvania State University and also a member of REHSEIS/CNRS at the University of Paris 7. She published two earlier monographs on the thought and mathematics of Descartes and Leibniz; her book *Representation and Productive*

Ambiguity in Mathematics and the Sciences appeared in 2007. She is the author of five books of poetry, most recently *Feuilles/Leaves* in collaboration with Farhad Ostovani. She has been an advisory editor for *The Hudson Review* since 1984 and frequently publishes in its pages.

Lauren Gunderson is a resident of New York City (formerly of Atlanta). A playwright, actor, poet, and story writer, she is at the intersection of science, history, feminism, and global humanism. Her plays have had many productions and won several prizes; some of them are collected in *Deepen the Mystery: Science and the South Onstage*, published with iUniverse. Her science-themed poetry appears in *Riffing on Strings* (ed. Sean Miller). Her short story, "Ascending Life," in this volume, won the Norumbega Short Fiction Award (as "Cancer/Dish").

Philip Holmes, born in England in 1945, moved to Cornell University and then in 1994 to Princeton University, where he is now Professor of Mechanics and Applied Mathematics and a member of the Neuroscience Institute. He works on nonlinear dynamical phenomena in solids, liquids, and biological systems. His wide-ranging investigations extend to planetary orbits (as in *Celestial Encounters*, jointly authored with Florin Diacu) and to insects (a current project is the study of how cockroaches can run so fast and straight). Among his honors, he is a Fellow of the American Academy of Arts and Sciences and an Honorary Member of the Hungarian Academy of Sciences. He has published four collections of poems, all with Anvil Press.

Alex Kasman is a mathematics PhD (Boston University, 1995), now an Associate Professor at the College of Charleston (SC). His scientific work extends to physics and biology. He has published a book of mathematically themed short stories, *Reality Conditions* (2005). Alex's website http://math.cofc.edu/kasman/MATHFICT is by far the most extensive resource on fiction dealing with mathematical themes, with over 650 entries.

Ellen Maddow is a founding member of the New York City troupe *The Talking Band* and has composed music for and performed in most of its productions. Works that she has written include *Painted Snake in a*

Painted Chair (2003 Obie Award) and five pieces about the avant-garde housewife Betty Suffer. Ellen received a 1996 McKnight National Playwriting Fellowship, the 1999 Frederick Loewe Award in Musical Theatre, and other awards.

Marjorie Wikler Senechal, a geometer and writer, is Louise Wolf Kahn Professor Emerita in Mathematics and History of Science and Technology at Smith College, where among other things she served as the founding Director of the Kahn Liberal Arts Institute. Her many books include *Quasicrystals and Geometry*, *Shaping Space* and—outside of mathematics—*Long Life to Your Children! a portrait of High Albania*, and *American Silk 1830–1930*. Marjorie is, with Chandler Davis, co-Editor-in-Chief of *The Mathematical Intelligencer*.

Manil Suri is Professor of Mathematics at the University of Maryland, Baltimore County, where he also holds the Elkins Professorship for the academic year 2007–2008. Manil is also a distinguished novelist, and received a Guggenheim Fellowship for fiction in 2004. His first novel, *The Death of Vishnu*, won the 2002 Barnes and Noble Discover Prize. His second novel, *The Age of Shiva*, appeared in 2008 to critical acclaim. These are the first two of (at least) a trilogy, he promises.

Randall Wedin is a chemist who has been a professional freelance science writer since 1991; he is also the primary parent of boys now aged 21, 18, 15, and 13. His interest in bridging the gap between science and humanism goes back to his undergraduate degree (1977) in Chemistry and English Literature, with an essay on using thermodynamics and crystallography as metaphors for analyzing literature. He worked for the American Chemical Society 1983–1981 in positions involving publishing and public communications. While there he initiated National Chemistry Week, a public outreach program, and received an award from the Public Relations Society of America for this. His essays have appeared in *Ruminator Review*, *The Chronicle of Higher Education*, *Chemical & Engineering News*, the *Star Tribune*, and elsewhere.

Paul Zimet is the Artistic Director of *The Talking Band*, a New York City company that creates original interdisciplinary works for theatre. He is also Professor Emeritus in the Department of Theatre at Smith College. Works he has written include *Imminence*, *Belize*, and *Star Messengers*, a musical on the birth of modern astronomy. Among his many awards are the John C. Lippmann New Frontiers Award and a 2003 Obie for directing *Painted Snake in a Painted Chair* by Ellen Maddow.

Jan Zwicky, a musician and a poet, also teaches in the Philosophy Department at the University of Victoria. She has published seven collections of poetry, including *Songs for Relinquishing the Earth*, which won Canada's Governor General's Award in 1999, and *Robinson's Crossing*, which won the Dorothy Livesay Prize in 2004. Her books on philosophy include *Lyric Philosophy* and *Wisdom & Metaphor*. She has also published widely on issues in music, poetry, and the environment.